TECHNOLOGY'S

CRUCIBLE

THE JAMES MARTIN BOOKS

- Application Development Without Programmers
- Communications Satellite Systems
- Computer Data-Base Organization, Second Edition
- Computer Networks and Distributed Processing: Software, Techniques, and Architecture
- Design and Strategy of Distributed Data Processing
- Design of Man-Computer Dialogues
- Design of Real Time Computer Systems
- An End User's Guide to Data Base
- Fourth-Generation Languages, Volume I: Principles
- Future Developments in Telecommunications, Second Edition
- Information Engineering
- An Information Systems Manifesto
- Introduction to Teleprocessing
- Managing the Data Base Environment
- Principles of Data-Base Management
- Programming Real-Time Computer Systems
- Recommended Diagramming Standards for Analysts and Programmers
- Security, Accuracy, and Privacy in Computer Systems
- Strategic Data Planning Methodologies
- Systems Analysis for Data Transmission
- System Design from Provably Correct Constructs
- Telecommunications and the Computer, Second Edition
- Telematic Society: A Challenge for Tomorrow
- Teleprocessing Network Organization
- Viewdata and the Information Society

with Carma McClure

- Action Diagrams: Clearly Structured Program Design
- Diagramming Techniques for Analysts and Programmers
- Software Maintenance: The Problem and Its Solutions
- Structured Techniques for Computing

with The ARBEN Group

- A Breakthrough in Making Computers Friendly: The Macintosh Computer
- Fourth-Generation Languages, Volume II: Representative Fourth-Generation Languages
- Fourth-Generation Languages, Volume III: 4GLs From IBM
- SNA: Principles of Systems Network Architecture
- VSAM: Services and Programming Techniques

with Adrian Norman

- The Computerized Society

TECHNOLOGY'S

CRUCIBLE

James Martin

An Exploration of the Explosive Impact of Technology on Society
During the Next Four Decades

Prentice-Hall, Inc.

Englewood Cliffs, New Jersey 07632

Library of Congress Cataloging-in-Publication Data

MARTIN, JAMES (date)
   Technology's crucible.

   1. Technology.   2. Twentieth century—Forecasts.
3. Twenty-first century—Forecasts.   I. Title.
T20.M348   1987   303.4'9   86–91492
ISBN 0–13–902024–1

Editorial/production supervision and
   interior design: Kathryn Gollin Marshak
Cover design: Whitman Studio
Manufacturing buyer: Gordon Osbourne

Printed in the United States of America

10   9   8   7   6   5   4   3   2   1

ISBN   0-13-902024-1   025

Prentice-Hall International (UK) Limited, *London*
Prentice-Hall of Australia Pty. Limited, *Sydney*
Prentice-Hall Canada Inc., *Toronto*
Prentice-Hall Hispanoamericana, S.A., *Mexico*
Prentice-Hall of India Private Limited, *New Delhi*
Prentice-Hall of Japan, Inc., *Tokyo*
Prentice-Hall of Southeast Asia Pte. Ltd., *Singapore*
Editora Prentice-Hall do Brasil, Ltda., *Rio de Janeiro*

TO CORINTHIA

CONTENTS

ACKNOWLEDGMENTS

Many versions of the material in this book were used for different purposes before the book was written in its current form. Some fictional events in the early drafts became reality too fast. A previous version contained a nuclear plant disaster remarkably like Chernobyl and a space shuttle explosion like that of the Challenger. An early television treatment written by the author in 1979 was restructured and added to by the British poet Leo Aylen. Small segments of Dr. Aylen's material are used in this book in a modified form. I wish to express my admiration for Dr. Aylen's articulate and compassionate views about society.

I am indebted to many people for suggestions and criticisms, particularly Dr. Carma McClure, Dr. Harvey Simpsohn, Ms. Caroline Coulter, and Dr. Leonard Kleinrock. To all these I wish to express my gratitude.

Original art was created for the book by Ms. Christine Phillips, who did botanical drawings in combination with architectural drawings by Mr. Duncan Coates. Ms. Jenny Walters created drawings of future life styles.

Art Sources

- Christine Phillips and Duncan Coates, Bermuda: pp. xvi, 4, 5, 6, 7, 8, 31, 34, 56, 57, 58, 84, 106, 152, 177, 178, 179, 204, 205, 206.
- U.S. Department of Defense, Russian Military Power: U.S. Government Printing Office, Washington, DC, 1985, pp. 128, 129.
- Jenny Walters, Cumbria, England: pp. ix, x, 32, 33, 82, 83, 180, 203.
- Fern Art, Cumbria, England: pp. 104, 105, 130.
- Paolo Soleri, Arcology: The City in the Image of Man: MIT Press, Cambridge, MA, 1969, pp. 37, 39, 42, 62, 113, 121: pp. 34, 106, 150, 151, cover.

DATE: December 31, 2019

TV NARRATOR: One of the most amazing factors when you look back at the history of technology is that the public in the late twentieth century had not a glimmering of what was really happening to them.

Earlier, it could not have been expected. But by the 1980s the dawn of machine intelligence was clear. If only they had recognized it then. Other factors that were so traumatically to change the world were clear too.

The interesting question is whether the course of history would have been different if the public had been able to comprehend the journey on which they had embarked.

TECHNOLOGY'S
CRUCIBLE

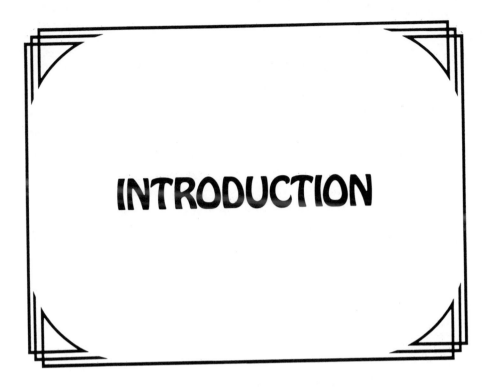

INTRODUCTION

This book is intended as a vehicle that might help you invent your own future.

As we look at the next 40 years, there are a number of things we can be reasonably sure about. First, human nature will not change. Shakespeare describes with astonishing skill human nature as we observe it today. We have the same power struggles, greed, love, kindness, jealousy, and treachery as in Shakespeare's day. It is a reasonable bet that it will be the same 40 years from now.

This same human nature will be set against a background of astonishingly different technology. We can be sure that the technology of the future will be no less spectacular than that described in this book. As always, there will be some surprises in technology that are not forecast here.

Assume, then, as you read about the future that the book has two things reasonably correct: human nature and the spectacular evolution of technology. Around those two aspects of the future we could weave a variety of images of society. You are invited to explore in your own mind the different future possibilities and to ask yourself,

"Given this technology and human nature, how can we build a better world?"

We are involved in a technological revolution of potentially greater impact than the industrial revolution. But unlike the industrial revolution, it is happening at devastating speed. It will leave no business, no industry, no institution untouched.

- Microelectronics
- Genetic engineering
- Chips doubling their capacity every year and a half
- Artificial intelligence chips
- Small silicon chips being replaced by large silicon wafers
- New technology chips of much higher speed
- Worldwide networks linking millions of computers
- The production cost of 100-kiloton nuclear bombs falling to $50,000
- Microbiology
- Robots
- Mass-production factories operating largely without workers
- Nuclear fusion
- "Star Wars" defense systems
- Optical fibers that can transmit the whole of Shakespeare in a quarter of a second
- Computers for artificial intelligence 10,000 times faster than those of today

This technology has shattering long-term potential. Forty years from now, society will have changed dramatically. New societal patterns are being forged in the crucible of high technology. Technology is changing exponentially, not linearly. The change in the next 40 years will be as great as that in the last 80 years.

Sometimes it seems that technology is moving so fast that it is out of control. Soon it will be moving much faster. The era of artificial intelligence has barely begun yet. Computers are being used to design ever-better computers. The explosive growth in computer power speeds up the development of genetics engineering, robotics, weapons systems, and all other technologies. How can the institutions of today's society—government, jobs, schools, the family—stand the strain of change?

Few people are thinking about the impact of future technology. Ours is an age without pragmatic philosophers. Most people who write about the social impact of technology write about past or present technology, not future technology. A new subject is needed in universities: the philosophy of technology. In university bookstores this subject, so vital to our future, is represented by a few negative diatribes against technology and by books that do not represent realistically what is likely to happen in practice.

It is the intent of this book to explore the impact of technology over the future lifetime of most readers in as graphic a fashion as possible. The book is written in the form of a television series, broadcast in the year A.D. 2019. Although it is set in the future, it is about the choices facing society today.

I have used the format of a television "treatment" in order to create a vivid image of the future and to achieve the maximum density of ideas in a book that most readers are likely to finish.

The series weaves a tapestry stretching from the industrial revolution to 2020. Some threads of the tapestry are remarkably constant for the entire period. The future can be made plausible by showing the continuity of human forces and behavior, set against the utterly spectacular changes in technology.

The motivations of the entrepreneurs and inventors of the industrial revolution are remarkably similar to the motivating forces today. The same will be true in the future. Similarly, today's would-be technology wreckers are remarkably similar to the nineteenth-century Luddites, who wrecked the first steam-driven weaving machines. They will have their counterparts in the future. So will the financiers, the politicians, the vested interests, the defense establishments.

Different programs in the series take different threads of constancy and show how the lessons of the past apply to the future. In exploring each of the threads, it is emphasized that as we enter the era of artificial intelligence and worldwide computer networks, the next ten years are absolutely pivotal.

This book does not forecast the future. It provides a vehicle for helping people think constructively about the future. A major theme is that we are not at the mercy of technology. We can channel it to our own ends. However, we may be at the mercy of our own human nature.

What is the purpose of it all? To build a more civilized world? A utopia? A golden age?

There are a few brief periods in human society that might be described as a golden age. What kind of golden age could we achieve if the world population is 8 billion and machines are ultra-intelligent? And what would endanger it?

If Western society made the right decisions over the next 20 years and the mass public understood the issues, there would be excellent prospects for a golden age of civilization such as the world has not yet seen, and within the lifetimes of most readers. If young people understood this and were determined to make it happen, our future would be brighter. For this reason I would like to see this subject matter filmed, used in schools, and broadcast everywhere.

7

Installment One

THE SORCERER'S APPRENTICE

PRESENT-DAY AUTHOR: *We are at the start of a technological revolution that will have greater impact on society than the industrial revolution. But unlike the industrial revolution, it is happening at devastating speed.*

A wave of technology is crashing upon us that will leave no business, no family, no industry, no institution untouched.

VIDEO: The following spoken list items are illustrated visually:

- Microelectronics growing faster and faster
- Chips doubling their number of components every 18 months
- Artificial intelligence chips
- Genetic engineering
- Silicon chips giving way to much larger silicon wafers
- Worldwide networks linking millions of computers
- Production costs of nuclear bombs falling below $50,000
- Microbiology
- Robots

- Mass-production factories operating largely without workers
- Nuclear fusion
- "Star Wars" defense systems
- Optical fibers that can transmit the whole of Shakespeare in a quarter of a second
- Computers for artificial intelligence 10,000 times faster than those of today

AUTHOR: Such technology has shattering long-term potential.

Toward the end of the lifetime of most of you, society will have changed, perhaps beyond recognition. The change rate of technology is speeding up.

I spend most of my working life with the professionals and executives who are making such things happen. They are changing the future of mankind perhaps faster than any group of people in history. Yet there is a remarkable reluctance to discuss that future or to attempt to invent the future. Conversations about the future tend to be about bits per second, software techniques, or future chips, not about civilization, quality of life, unemployment, control of future weapons systems, or changes in the democratic process—and certainly not about the meaning and purpose of it all.

On the other hand, the persons outside of technology who are involved in society's fabric, the lawyers, teachers, bankers, politicians, writers, sociologists and managers, seem to have little comprehension of future technology or of how technology might change their institutions. That their institutions are going to change there is no doubt. Many of these influential people seem to regard the discussion of technology as something not quite respectable, just as a nineteenth-century gentleman would not have talked about the plumbing.

Television, by far the most influential communication medium in history, gives an astonishingly distorted picture of future technology and its impact. It is either portrayed in science fiction terms with space adventures and humanoid robots, or else as a negative force with computers destroying privacy, dehumanizing work, and making possible the black vision of George Orwell. The future of technology is likely to be entirely different. It *will* drastically change jobs and working patterns. It will change war and politics. It will change life styles, change cities, change the arts, theater, and leisure. It raises subtle issues not directly connected with technology, such as how people spend their time when factories and offices are fully automated.

The technology of the next 40 years is the most potent force for change. Certain aspects of it are highly alarming and dangerous. Overall, technology gives us the capability to build a much better society, but we may blow our chances of that and blunder into undesirable worlds by accident if we do not discuss the future. Particularly dangerous is to view such a potent force through the grossly distorting lens of today's television.

We can make choices, and we need to understand those choices. Historians of the future will look back on the next ten years as being a pivotal decade in changing the course of society's evolution.

The series has been filmed as though it were made in the year A.D. 2019. It looks back on the events and driving forces of technology. It is not intended to be science fiction but science philosophy. If it makes you argue about your future, it will have served its purpose.

The programs emphasize that they do not forecast the future. They provide a vehicle for helping people think constructively about the future. A major theme is that we are not at the mercy of technology. We can channel it to our own ends.

VIDEO: A scene-changing sequence is used to slide from the present to A.D. 2019: a rushing sensation like flying very low at extreme speed through a mountainous valley, but the earth and hills look like endless etched microelectronic circuitry. Electronic music is used, evocative of time travel.

TITLING: AUSTRALIA
December 31, 2019

VIDEO: A large church of ultramodern architecture has towers spanned by a lattice holding many bells, large and not so large. The bells boom out across the city.

We see punts on a stretch of river overhung by willows. Women as well as men are punting. On the banks, women carry yellow parasols to protect them from the strong sun. There are many flowers, the unusual-looking flowers of Western Australia, banksia, sun orchids, and waratah. The punts drift under an ornate bridge covered in sculpture. The images of life are relaxed and civilized.

The camera tracks through a section of city that hardly looks like a city. Most of the buildings are partially underground, their windows showing through grassy banks. Gardens with people walking stretch over the buildings. The camera closes in on a doorway that has a large sign: WORLDCOM.

A television studio. Unmanned cameras move downward. They are on 20-foot arms with joints that extend from a robust ceiling track. Other arms hold lights. A camera closes in on a dignified seated figure. Its lens rotates automatically with a whirring sound. The robot lights adjust their position. The figure is middle-aged and sun-tanned. His elegantly cut suit is the latest fashion for the 2020s. He sits at a desk with a transparent top; computer screens are visible beneath its surface.

> *Head of Project Jacob*: We are about to begin a new decade, a decade in which we are challenged to extend our mastery of technology by a quantum leap.
>
> The benefits from Project Jacob will affect all humankind. It is a great step outward in creating a planet that can be safe and clean, where all human beings can develop their capabilities to the full. We will look back on past decades with curiosity and affection. Some people crave the past. But we must understand that most human beings were prevented from growth; they were stunted by lack of opportunity for education, by lack of powerful mental tools, by the need to slave for the prime hours of their life at drudgery jobs, by lack of nourishment both mental and physical, and by lack of vision.
>
> The time has come to attack the ignorance and drudgery that remain in the world and to unlock the creativity that exists in all people.
>
> Project Jacob is expensive. No one country is likely to pay for it by itself. No one corporation could afford to. It is a project in which many governments and corporations must cooperate.

VIDEO: The camera cuts to a professor at his university. He can see the previous speaker in an electronic window.

> *Professor*: Project Jacob is outrageous. The very suggestion that we should spend public funds on it is a perversion of what it means to be civilized. We should show the public in detail what the same amount of money would buy if we spent it on the arts, drama, worldwide videodisk libraries, Michelin three-star restaurants, . . .

NARRATOR OF THE SERIES, CORINTHIA: The debate over Project Jacob is perhaps the most heated debate we have ever seen on a worldwide scale over a project in technology. The advocates of technology are adamant that Project Jacob must be built. There is perhaps no project in history that has had such determined advocacy by the technology-minded. On the other hand, many members of the public are outraged that such a massive expenditure should ever be taken seriously, especially when so many other causes need money so badly.

The debate relates to civilization itself. What does it mean to be civilized? Technology is unlocking vast riches. How should such riches be spent? Can we create one of the golden ages of civilization? What do we mean by a golden age?

> *VIDEO:* The narrator, Corinthia, is a dynamic, attractive woman in her forties. She conveys both intelligence and enthusiasm. As she talks, we see the kind of place she lives in: her semi-underground house, the surprising elegance and beauty of her life style; we see her garden robots doing various jobs in the background.

NARRATOR: In discussing civilization it is extremely important to distinguish between ends and means. Most persons are preoccupied with *means*. Technology, law, management, politics, programming, corporations, the civil service, and making money are all *means*, not *ends* in themselves. Ends relate to how we enjoy the fruits of such labor or moneymaking. When we lose the perspective to concentrate on ends, the means take over. We perform work for its own sake and create more work. Civil servants multiply their bureaucracy, destroying not only their own pleasure but other people's pleasure also. Lawyers think up ever more byzantine and destructive machinations.

A highly civilized society concentrates on the ultimate purpose of our labors and minimizes the drudgery needed to achieve the purpose. It refines as fully as it can the pleasures that civilization can achieve.

Technology has an immensely important role to play in achieving this objective. It can remove drudgery. It can create wealth. It gives us superb hi-fi systems and projection TV. It gives us access to knowledge and computerized logic. It gives us mobility. It gives us the machinery of filmmaking and theater.

However, like the other *means*, it tends to take on a life of its own unrelated to the *ends* of being civilized. When technology becomes an end in itself, it is pointless. We need to regard technology as a tool to achieve a greater goal.

A striking aspect of the history of technology is that new technologies destined to change society dramatically have always taken us by surprise. There has always been a reluctance to anticipate or believe the implications of powerful technology. The industrial revolution was such a fundamental change in human history that it could not have been anticipated. By the twentieth century the early warnings of new technology should have been heeded.

As we look back, one of the most pivotal periods was 1985 to 1995. Microchips were cheap, petroleum destined to become expensive. Robots spread rapidly in industry. Silicon chips gave way to much larger silicon wafers. Vast artificial intelligence projects were created, expert systems flooded the marketplace, research began on biological chips, genetic engineering became a major growth industry, and computer networks spread to every corner of the industrial world.

It should have been obvious as far back as the 1970s that the technology being unlocked then was destined to change society completely.

Very few people thought about it then. Very few people thought about what sort of world they wanted to build.

A great threshold in our development was reached when machines became much more capable than humans for functions requiring complex knowledge and inference processing, and vast storehouses of such capability began to grow and spread. This was the dawn of machine intelligence.

One of the amazing factors when you look back from today is that the public at large in the 1980s, on the brink of this great threshold, had no clue to what was really happening to them. Earlier, it could have been expected. But by the 1980s, the dawn of machine intelligence was clear. If only they had recognized it then. Other factors that were to change the world dramatically were clear too.

Technology by itself would have been a bore to many people. But its far-ranging byproducts were not. It freed people from drudgery so that they could play or rethink and concentrate on the arts, family, environment, or quality of life. It gave tools for culture. It freed the mind to create and to rebel, to build and to destroy, to argue and to reshuffle the cards.

VIDEO: Montage of orchestras, gardens, motorcycle thugs, dinner parties, theater, park orators, church, elegant flirtation.

NARRATOR: We can look back at America and Europe now and ask the question, How would society have evolved differently if the public at large had understood what was happening?

Surely the course of history would have been different if the public had been able to comprehend the journey on which they had embarked. By 1980 the public at large had become aware of the microchip—a tiny silicon-based circuit with thousands of components that could be a computer, a memory unit, a transmitter, or a logic circuit. But few people realized the incredibly rapid rate of improvement of chips.

The number of components on a chip doubled every year for many years.

VIDEO: Pictures of chips becoming more and more complex.

ANIMATION: It is sometimes difficult to grasp the effects of constant doubling. Suppose that in 1959, when the first transistor was printed on silicon, a patch of seaweed in the Pacific Ocean measured one foot across. The seaweed patch doubled in size every year just as chips have doubled their number of components. By 1964 it would have measured 32 feet across; 1964 chips contained 32 components. By 1970 it would be 2000 feet across. By 1984 it would have choked the entire Pacific.

NARRATOR: Microelectronics did not stop growing in the 1980s. The number of components on a chip doubled every year and a half. In the 1990s the silicon wafer supplemented the chip. A wafer, at first 3 inches across, then longer, contained hundreds of millions of components. It could not be manufactured without defects, so the key to making such wafers was to design them so that faulty components on the wafer could be bypassed automatically. Silicon wafers, like chips, would be mass-produced like newsprint. Once the manufacturing problems of making an integrated wafer were solved, wafers increased in both density and design sophistication at a furious rate, quadrupling in capacity every three years. The first wafers had 10 million components, then soon had 100 million. By the year 2000 wafers were sold with a billion components. Soon after, optical wafers had 10 billion, then 40 billion components with much memory and many computers operating in parallel on the same wafer. Many such wafers, especially memory wafers, were packaged into one device.

Factory-like machines were used in orbit to make microelectronic components far closer to perfect in the total vacuum of space.

VIDEO: A spacecraft with solar sails 200 meters long, for generating electricity. The satellite has a body 40 meters long and a small cluster of antennas, the two largest of which are 3 meters across.

The narrator holds up a thin silvery object the size of a 20-centimeter ruler glistening with rainbow-like colors (like a videodisk).

NARRATOR: This mass-produced wafer contains almost 20 billion components. It contains 64,000 computers, each with 4000 bits of memory. These can be interconnected in any configuration. Thirty years ago this power required one of the world's largest supercomputers.

> *VIDEO*: The operator picks up a 5-centimeter object of similar appear
> ance.

NARRATOR: This is much less expensive. It costs less than a digital
wristwatch cost in 1980. To have performed the same computations
as this in 1960 would have needed a machine larger than Los Angeles.

When the public first became excited about microchips around
1980, little did they know that electronic circuits were going to become
a million times more capable for the same price.

> *ANIMATION*: A major change in computing occurred when it became
> common to use many processors in parallel. For the first half century
> of computing, almost all computer operations were performed as a
> sequence of instructions on one machine. We used exclusively sequen
> tial programs that performed a single step at a time.

NARRATOR: In the late 1980s microcomputers became very inexpensive. By the late 1990s thousands of microprocessors could coexist
on one mass-produced wafer. The cost became less than 10 cents
per processor. Personal computers by the year 2000 were described
in terms of kiloprocessors (thousands of processors) rather than kilobytes (thousands of bytes of memory). It became necessary to think
of computing in a new way. What applications and what types of
operations could be carried out on highly parallel machines? Parallel
machines were excellent for searching and merging large amounts
of data, looking for rules to apply, tackling problems in which many
logical inferences could be handled simultaneously, processing images,
generating elaborate displays, processing digital television, complex
simulation and design, and, particularly, artificial intelligence applications.

Artificial intelligence began to find profitable applications in the
1980s. It boomed into a large industry in the 1990s.

> *VIDEO*: Montage of headlines from the financial press and Wall Street
> reports about the profitability of artificial intelligence (AI). Images of
> robots, space stations, hydroponic gardens, bioengineering plants, and
> children excitedly interacting with computer screens.

NARRATOR: The first four generations of computers were modeled
after the left brain. They were sequential, logical, algorithmic, quantitative, and data-based. Fifth and later generations were modeled after
the right brain also: intuitive, qualitative, holistic, artistic, heuristic,
knowledge-based—and parallel.

Artificial intelligence was slower to evolve than expected at first because in the 1960s we underestimated the complexity of our brain and its thought processes. But slowly the work paid off. By 1980 there were cheap chess-playing machines that only a grand master could beat; there were industrial robots, and diagnostic programs could beat doctors at their own diagnoses in certain specialized areas of medicine. In the late 1980s expert systems were built that packaged many different forms of human expertise.

From these small beginnings a vast industry and awesome machine capability was to grow.

By the 1980s it should have been clear that, for numerous specialized functions, machines would become more skilled than humans. When a machine first beat a human at chess, we begrudgingly thought of it as having a form of intelligence. Today we see millions of different forms of this machine intelligence.

ANIMATION: Networks covering the globe and machines multiplying in vast numbers.

NARRATOR: It was common to think of machine intelligence as being in one machine in one room. We should have realized that it would be much more than that. There are millions of computers in worldwide communication. They have access to vast knowledge bases and can intercommunicate in a fraction of a second. The workstation in every home and office has access to this network.

When expert systems and computer-aided design became very powerful, it became possible to design machines and software much faster with a much higher level of automation. The fifth generation of computers greatly enhanced the ability to design the sixth generation. The sixth generation featured phenomenal machines that produced extremely advanced designs very quickly. A chain reaction built up. Machines designed machines. Intelligent systems learned at electronic speed. Knowledge bases became self-feeding.

Particularly important in the evolution of computing was the automation of programming. Before the 1980s computer programs had to be written a tiny step at a time. It was like telling a centipede every limb movement it must make in order to climb an apple tree.

VIDEO: Slow-motion close-up shot of a centipede on irregular twigs.

NARRATOR: In the 1980s application generators—programs that could write other programs—came into use. These improved steadily, and much of the burden of ordinary programming was removed. In the 1990s complex logic was created graphically at the screen of a workstation. The computer helped to check that the logic was correct, linking it to formal descriptions of data, and generated program code.

VIDEO: A montage of use of intricate color graphics showing the design of data and programs in multiple windows of a computer screen.

NARRATOR: As the process of programming was automated, a chain reaction began. Programs created other programs, and they steadily became better at it. Computers were employed to design other computers, and they improved steadily. The rate of improvement increased with each step.

The machines acted as an accumulator for human knowledge, the world knowledge bases becoming more and more comprehensive all the time.

Industrial robots first came into use in the late 1970s. Ever since, this has been one of the world's top growth industries, with new models of robots making the previous ones obsolete at a furious rate.

VIDEO: Montage of robots welding, moving parts, screwing bolts: medium-sized robots in car factories, giant robots in shipyards, robots moving steel girders, tiny robots assembling Swiss watch parts.

NARRATOR: The public's concept of a robot used to be completely different from the machines that actually became useful and commercially viable.

VIDEO: Sequences of robots from old movies, starting with 1920s and 1930s robots and evolving to the *Star Wars* robots. End with a sequence from the film *Saturn III* showing a robot accidentally programmed to lust after Farrah Fawcett.

NARRATOR: The first major use of robots was on production lines. These machines were not intelligent at first, but they greatly improved industrial productivity. The Japanese employed them in vast numbers and flooded the world with goods that were cheap yet highly reliable because fewer human mistakes were made. Countries without this form of automation became increasingly vulnerable to foreign competition.

The robot production lines spread and became ever more versatile. Consumer goods dropped and dropped in price.

The robots needed better eyes and other sensors, as well as more logic so that they could recognize objects and take more intelligent actions. A human has two eyes near the top. Is that the best place for eyes? Often not. A production robot can make good use of eyes in its fingertips.

VIDEO: We see a robot claw with three fingers. There is a tiny, hard lens in each. The claw picks up a complex object.

NARRATOR: The human has one brain. Robots often have many tiny brains for different purposes. This machine has a microcomputer built into its hand.

VIDEO: A human hand holds a robot claw and points to where its microcomputer is located.

NARRATOR: The challenge became to redesign the human being— to redesign the fabricator. You could have eyes wherever you want them, and many hands. You could use sensors that humans don't have, like magnetic sensors, X-ray sensors, motion detectors, or precision lasers. You could have an arm with many joints capable of welding where humans can't reach. You could have fingers the size of ant legs or big enough to shift many tons.

VIDEO: A diversity of future robot fabricators, like an intricate dance of insect-like parts.

NARRATOR: A one-armed robot, like a one-armed man, was limited in its capability. Two-armed robots were difficult to build in the 1980s because control mechanisms were needed to coordinate the arm movements. The arms could easily collide or jam each other, especially when they were holding components. Mathematical techniques were created for coordinating robot arms.

VIDEO: A two-arm robot attaching the eyes to a teddy bear.

NARRATOR: Once two arms were coordinated successfully, robots that used numerous arms when necessary were built. Communicating microcomputers controlled the robot limbs. The mathematics of robot limb coordination became an important subject in universities and research departments everywhere. Animal limb motion was extensively filmed, studied, and simulated mathematically. Wild creatures have beautiful and complex limb control that at first was exceedingly difficult to replicate with microbotics.

VIDEO: A robot clumsily walking on two legs. The legs of a runner in slow motion. The runner hurdles a fence. The computer screen simulation of the runner's limb motion. A gardening robot on four tall legs, stepping high to avoid plants. A close-up of a centipede's legs moving. A research machine with 20 legs like a centipede. A cheetah racing at 70 miles per hour nudges its fleeing prey; the antelope bowls over and the cheetah clutches its throat. A computer screen simulation of the cheetah's legs. A four-legged research robot running through grass. Scientists analyze its motion on a screen and watch a computer comparison of the cheetah and the machine.

NARRATOR: While much was learned in the mathematics of micro-botics from the study of animal limb motion, limbs were soon devised that were controlled in ways unknown in nature. Hands with six or ten fingers were used in jewelry cutting. Limbs were built with eyes and microcomputers in the limb itself.

VIDEO: A flight simulator pod for a jet fighter with six 20-foot legs tilting the simulator sharply.

A robot on six legs stepping across a rock-strewn plain on Mars; each leg has three lenses looking at the ground in different directions.

A giant excavator with large cams lifting its eight big flat feet.

NARRATOR: One of the largest-growing segments of the toy indus-try became microbotic animals. From Japan they poured into the world markets in all shapes and sizes. With or without radio control, the antics they performed became steadily wilder and more elegant. The early robotic toys moved slowly and crudely. Twenty years later they had full limb control. Ten years after that racing life-sized toy greyhounds became a billion-dollar fad. Toy cheetahs could chase and catch their prey almost like real cheetahs. A galaxy of science fiction creatures made toyshop windows a major street entertainment.

VIDEO: A toyshop in 2019: the windows of F.A.O. Schwartz on New York's Fifth Avenue at Christmas.

NARRATOR: As the fabricator changed, so it became desirable to change the products it made. The challenge to designers was enor-mous. Almost every mechanical product became redesigned for robot production. All manner of new designs were possible. We acquired intricate machines that could never have been built with human meth-ods.

Robots dropped in cost, rapidly, and vast markets opened up for domestic robots. Mowing the lawn or vacuum cleaning was drudg-ery on which human beings had to waste their time only 30 years ago.

VIDEO: A robot cleaner noses its way around a chair; another opens a cupboard and goes inside.

NARRATOR: To take full advantage of household automation it became necessary to redesign both the house and the garden. Land-scape automation became a booming industry.

VIDEO: A woodland scene with flowering shrubs and a finely mowed grassy path winding through the shrubs, its edge trimmed immacu-lately. We see a six-inch robot mower cutting the grass.

NARRATOR: Once robot cookers were widespread, the big market became their software. The world's best chefs were hired to create gourmet recipes that a robot could produce. The necessary sauces and ingredients were packaged along with the software.

VIDEO: A cooking machine moving packages of food to its work-stations. It slides a pan over a heater, and a stirrer slowly lowers into it. Drops of a red liquid drip into the pan.

NARRATOR: Mechanical goods dropped in cost because of robot mass production, but they could not drop indefinitely. The cost of computer memory, computer power, and transmission, on the other hand, has dropped in cost for 80 years and will continue to drop. In 1950 $100,000 would have bought only the tiniest computer memory, of very little practical use; by 1960 the same sum bought 4000 characters of core storage; by 1980, 5 million characters on chips; by the year 2000, 5 billion characters on a silicon wafer; now a trillion characters with molecular electronics. Similar cost reductions applied to processing power and to data transmission. Printers, screens, and keyboards could not drop in cost to such an extent.

Anything to do with the storage or manipulation of *bits* dropped in cost. A bit has no size or weight. It could be as small as molecular physics would allow and could travel at the speed of light. It is conceptual rather than physical. You cannot touch it. The techniques constantly improved, and the devices packed more and more bits in a tiny space and were mass-produced in ever increasing quantities.

The total number of digital functions in society that manipulated, stored, or transmitted bits grew at a great rate because the rapidly increasing power of digital devices was combined with an even more rapid growth in mass production. The number of wafers increased, and the power of each wafer increased, both at a furious pace.

This geometric growth did not abate for the half century from 1969 to 2019. Society was flooded with digital functions mass-produced like newsprint, finding their way into every type of gadget, every nook and cranny of every institution.

The geometric growth in ability to handle bits of information could not be matched in the growth of mechanical goods, but in one other area it *was* matched: molecular biology.

NARRATOR: (over video) All living things are composed of *cells*. A cell is as intricate and complex as a chemical plant. It takes in raw materials, such as oxygen and food compounds, and produces useful products and waste, which must be disposed of. Like a chemical plant, the cell is controlled by a complex set of operating instructions. These instructions guide and vary the activities of the cell. In a chemical plant the instructions come from a computer program. In a cell they also come from a type of computer program—the information encoded in the *genes*.

Cells provide nature's form of mass production. A cell is cunningly designed to reproduce itself.

VIDEO: Speeded-up microscope shots of complex cells reproducing themselves.

NARRATOR: Under certain circumstances cells reproduce themselves at a rapid rate. Cells that we use, such as *E. coli*, reproduce themselves about once every 20 minutes under appropriate conditions.

ANIMATION: One such cell in an hour becomes 8 cells; in two hours, 64 cells; in ten hours, a billion cells; and after a day of reproduction we have 4 thousand billion billion cells.

NARRATOR: Again the analogy of seaweed growing in the Pacific applies. The one-foot patch of weed doubling, redoubling, and doubling again would choke the entire Pacific not after 25 years but after 9 hours if it grew at the rate of *E. coli* cell reproduction.

VIDEO: A repeat of the sequence showing the weed patch doubling intercut with microscope shots of cells reproducing.

NARRATOR: By making cells reproduce we can mass-produce biochemicals. The biochemical industry searches for the best producers in nature, adapts them with appropriate mutations, and grows them in controlled conditions.

VIDEO: Footage inside biochemical plants showing vats of chemicals bubbling, mazes of pipes, and instruments.

NARRATOR: The original production of penicillin culture in the 1940s produced four units per tank of growing mold. After a search for the best natural producers, production was increased to 100 units per tank in the 1960s. With mutation and selection this was pushed to 1000 units per tank in the 1970s, 10,000 units per tank in the 1980s, and 100,000 units per tank today—nature's mass production.

The biochemical industry took a giant step forward when we learned how to change the program that controls the behavior of cells. We are not nearly clever enough yet to write the program controlling a cell and make our program function. What we can do is take pieces of functioning programs and splice them together to make different programs. We call this *gene splicing*.

In the 1980s, a new industry emerged to do this—the genetics engineering industry. Most of the new genetics engineering companies were in Silicon Valley, Tokyo, and other areas close to the burgeoning microelectronics companies.

> *VIDEO*: A montage of headlines from the financial press and Wall Street reports showing the financial fever over genetics engineering companies.

NARRATOR: By 2013 Cybergen was the world's second largest corporation.

A computer program is a collection of bits, each bit representing a zero or a one. In the early days programs were stored on paper tape where a one was a punched hole and a zero the absence of a hole. A gene stores a program in a somewhat similar fashion, except that instead of being binary the code is quarternary. The program is encoded as a string of submolecules known as *bases*. There are four types of submolecules: A, C, G, and T, standing for adenine, cytosine, guanine, and thymine.

> *VIDEO*: A paper tape containing a computer program moves horizontally across the screen. Below, moving at the same speed, is a representation of a DNA molecule with submolecules labeled A, C, G, and T in a random sequence.

NARRATOR: These are the letters *TRF* in computer code:

00101010 01001010 01100010

In our body, TRF is secreted by the hypothalamus of the brain. It is the master hormone regulating the action of the thyroid gland. This is its genetic code: GAGCATCCT.

> *VIDEO*: The two codes are together on the screen.

NARRATOR: Perhaps one of the most extraordinary facts about genetics is that this program code stored in DNA molecules is universal. It applies to insects, plants, giraffes, cockroaches, fish, bacteria, *and humans*. When we learned to tamper with DNA, we acquired the ability in principle to tamper with all of nature.

Splicing together genes in the laboratory, we have created new antibiotics, enzymes, hormones, vaccines, contraceptives, and agents that fight cancer. We can change cell information to create modified species of plants. We have produced better sheep and beef cattle. We can cure some inherited diseases. We have improved the intelligence of some animals and know that it is within our grasp to do that with humans. Today it is easier to work with DNA than with any other macromolecule.

To create new genes, the following steps are performed:

ANIMATION:

1. Lengths of DNA are cut from different sources. Restriction enzymes can cleave a long strand of DNA where a particular code sequence occurs, like cutting a computer program where a particular character sequence is found.
2. Separate strands of DNA are allowed to join together.
3. Recombinant plasmids (small loops carrying three or four genes each) are taken up by *E. coli* cells with their walls made permeable.
4. Only a tiny fraction of the resulting bacteria would contain the recombinant genes. The remainder are selectively killed with antibiotics.
5. The surviving bacteria, containing the new genes, multiply, forming spots or colonies on nutrient plates, each colony containing millions of identical cells.

NARRATOR: Biochemical development went through many stages: first, the search to find the best bacteria in nature for the required purpose; second, the introduction of mutations by chemicals or radiation, hoping that a few out of many mutations would prove valuable; third, gene splicing largely by trial and error, hoping that a few of the many genes produced would be useful; fourth, gene splicing planned with complex computation in a more calculated search for new genes.

Some of the most valuable biochemicals exist in nature in quantities too minute to be useful. One example is interferon. When a virus enters a cell in the human body, the cell secretes interferon, which passes quickly to other cells and stimulates them to produce appropriate antibodies to defend against the virus. Interferon thus acts as a powerful antibiotic against a wide variety of viruses. Interferon exists in the body in such minuscule amounts that it could not be extracted economically for commercial use. Recombinant DNA, however, allowed it to be attached to a fast-multiplying bacterium and mass-produced in large quantities. Many valuable biochemicals are mass-produced in a similar way.

A particularly important substance is referred to by the same initials as artificial intelligence, AI. AI (which stands for angiogenesis inhibitor) has been a major weapon in the long defeat of cancerous tumors. When a cancer tumor begins its growth, it has no blood vessels. It cannot grow large or dangerous without blood vessels. Normally, a tumor secretes a substance called TAF (tumor angiogenesis factor), which causes nearby capillaries to branch and grow, feeding blood to the tumor. The tumor can then grow and become malignant. AI blocks TAF and prevents the tumor from growing until it withers and dies. AI is mass-produced with gene splicing. Cancer, like polio and smallpox, has been largely conquered.

In the late 1970s genes were first introduced into mice by injecting the DNA into mouse ova at the moment of conception. Identical twin mice (clones) were produced, and it was possible to breed mice with particular characteristics.

VIDEO: Shots of identical mice scurrying, hypodermic needles in petri dishes, microscope shots of cells dividing, laboratory rabbits, a microscope shot of a needle piercing the wall of a cell, a human baby laughing.

NARRATOR: Around the same time the first human "test-tube babies" were created. The mothers who gave birth to these babies were generally delighted and looked after their babies with special care knowing that they could not have conceived a child the normal way. Most test-tube babies grew up to be fine people. It was many years before *gene-transplant* babies were created.

VIDEO: A sequence of headlines and magazine covers hailing gene-transplant babies as a sensational advance in science or condemning them as tampering with God's will. A mother with a happy child.

NARRATOR: Like heart transplants, gene transplants were used only when all else failed—when both parents carried genes for an inherited disease, for example. The disorder was prevented not only in the new baby but also in that person's children. Amniocentesis, a procedure that studied cells to check a fetus for possible genetic problems, became a common procedure in affluent societies.

Genetic manipulation was carried much further with animals than with humans. A type of dog that gained great fashionable popularity in the 1990s was the Chinese Shar-pei. This dog had a beautiful wrinkly skin that seemed three sizes too large.

VIDEO: Shots of Shar-pei dogs: first amiable dogs, their owners petting and stretching the dog's strange skin, then a snarling dog.

NARRATOR: These dogs commanded high prices and so were bred at a furious rate. The Shar-pei had been a rare dog, and the market demand resulted in severe overbreeding. The dogs became unpredictable and sometimes vicious. The overbreeding caused genetic defects, so the Shar-pei became a target for gene improvement experiments. It became highly profitable for the breeding kennels, mostly in China, to eliminate the gene defects, and the Shar-pei became the first creature to which gene manipulation was applied on a large scale. It became possible to breed Shar-peis that were affectionate and highly intelligent. Raising the intelligence of an entire species had not been done before and gave rise to speculation about the extent to which China was doing similar experiments with humans.

The combined effect of electronic technology, robots, and genetic technology was to change everything in society. The structure of corporations changed because the flows and processing of information changed. Decision making by knowledge workers everywhere was done with the aid of machines, and there were few workers other than knowledge workers. Factories operated with almost no people, mass-producing highly complex goods. Interactive television changed the nature of democracy, and electronic surveillance permitted abuse of freedom. Drugs of all types became cheap and widespread—mood-changing drugs, hallucinatory drugs, tranquilizers, energizers, erotic drugs. Video programs became freely available from libraries—the best, the worst, the most intelligent, the most obscene. Telecommunications networks enabled institutions to be worldwide. A wired society grew into a wired world. The genetic engineers invented new crops, new fertilizers, new drugs. They tampered with animals and argued about how far they should go in tampering with man.

> *VIDEO*: A montage of unfamiliar images of different beef cattle, hypersonic planes, mass festivals of samba dancing with laser beams, spider-like robots. The images are on cards which become rapidly shuffled. Dissolve to a European stained-glass window with angels.

NARRATOR: When the cards are shuffled so drastically, society cannot operate adequately without guiding principles. The churches have lost their nineteenth-century authority. Modern science, while supporting the concept of God, has questioned much of the established religious lore.

A sense of values supported by the churches, schools, and television is essential for adequate functioning of a society. This sense of values needs to be clearly written down, articulated, and taught. It needs to be a guiding set of principles for the media. Institutions that violate the value system need to suffer the full focus of media and public attention. Much of the chaos associated with the rise of technology in the last half of the twentieth century was caused by the breakdown of society's traditional value systems.

The value systems of the world's religions have much in common. A code was drawn up in the 1980s extracting 17 guidelines for behavior that are largely common among different religions. A number of schools adopted and taught this, but it did not pervade much of society. It was clear that countries with a firm adherence to such a value system for children were better able to cope with the uncharted seas of technology.

We need more, however, than the value systems of the old religions. A vital question today is, What is the purpose of it all? There is no point in unlocking more and more powerful technology unless we build a better world. What sort of a world do we want?

Society today is rather like the sorcerer's apprentice. We are unlocking technology that to an earlier era would have appeared magical, but we are naive about what to do with it. It may get so out of hand that we have to find ways to control it.

VIDEO: (over the orchestral suite "Sorcerer's Appentice" by Paul Dukas) A hypodermic needle in a petri dish. A microscopic shot of a needle piercing the walls of a cell. Slow dissolve to an image of Aladdin's lamp and its genie. Dissolve to a drug peddler in a disco, spaced-out faces, newsreel footage of people in Bhopal, India, after the Union Carbide chemical plant disaster in November 1984, footage of the Chernobyl nuclear disaster. Slow dissolve to a red-hot crucible with vapor rising from it.

NARRATOR: Technology has given us the capability to work miracles. This should have been clear in the 1980s when we began to see the potential of artificial intelligence, robots, and genetic engineering. The vital question then becomes, What miracles do you want?

The French word for automation is *génie*, the same as the magical genie that comes from Aladdin's lamp—a word similar to *gene*, the program controlling the behavior of cells.

The world's childhood fairy tales are full of stories about a person being able to ask for miraculous wishes, have them magically satisfied, and find that the outcome was disastrous. When technology works miracles, we can likewise be surprised at the outcome.

To select how our institutions function in a world of technological miracles and extreme change, society needs rules that relate to the value system. The rules act rather like genes in setting certain behavior patterns in a highly complex organism. With damaged genes it is likely to go badly astray.

VIDEO: Animation representing damaged genes, merged to scenes of police quelling violence with fire hoses. A montage of screaming, passionate faces filmed during violent riots.

NARRATOR: The sense of values for society must operate at four levels. The first and most fundamental is basic ethics, the value system expressed by the world's religions.

VIDEO: A flat building block, the first of a pile of four, labeled "Basic Ethics" (Fig. 1.1).

Figure 1.1

NARRATOR: When the teaching or acceptance of basic ethics breaks down, we see a major increase in crime, drug abuse, divorce, unemployment, and general unpleasantness in society.

The second level is the values expressed in political constitutions, of which the American Constitution is one of the best examples.

VIDEO: A second building block on top of the first. It is labeled "Political Constitution" (Fig. 1.2).

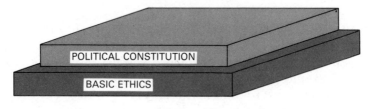

Figure 1.2

NARRATOR: General knowledge of this constitution by the public and its acceptance by the media resulted in outrage after the Watergate affair and the ouster of a president who abused the constitutional rules.

NEWSREEL: President Nixon leaving office.

NARRATOR: The Constitution has helped preserve the high level of freedom that exists in America, and, related to it, the high level of entreprenuerialism. This has been true throughout an era when the government and corporations have had access to a technology for extreme surveillance and control. George Orwell's *Nineteen Eighty-four* has not come true in most Western countries, although the technology to make it happen is immensely more powerful than Orwell envisioned.

VIDEO: The third block, labeled "Welfare Systems," is placed on the others (Fig. 1.3).

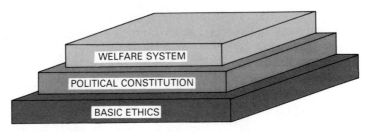

Figure 1.3

NARRATOR: A society of reasonable wealth needs a welfare system to ensure that its citizens have good health care, good education, and basic services, and do not starve if they are unemployed.

The top level is more subtle. A society may avoid excessive crime, drug abuse, political oppression, and surveillance. Its citizens may have total freedom and substantial wealth through automated industries. They may be supported by a welfare system. But it could still be a barbaric society. The top level of values relates to what it means to be civilized. What are the characteristics of great civilizations?

VIDEO: The fourth building block, labeled "Civilization Values," is placed on the others (Fig. 1.4).

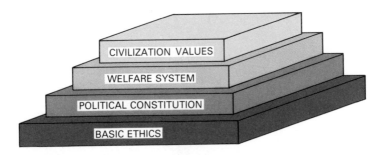

Figure 1.4

NARRATOR: Certain brief periods in human history might be described as a golden age—an age when humankind appeared truly civilized: the Athenian civilization from the Battle of Marathon in 480 B.C. to the death of Alexander in 323 B.C., Italy in the fifteenth and sixteenth centuries, France from 1653 to 1788. If we ask what made these brief eras so special, the answer relates to the same sense of values that permeated their privileged classes.

Ours is an age of much greater material wealth, an age when we can work miracles with technology. It is an age when we would want to apply the term *civilization* not just to a privileged class but to all of society—a society in which men and women are equal and only machines are slaves.

Could we attain a golden age like the Athens of Pericles or the France of Voltaire set against a background of modern technology? If we cannot, what is the point of further technological advancement?

What is the best way to spend our technology-generated wealth? On yet more technology, like Project Jacob, or on changing the values of a society to make it more civilized? And who can make such a choice? What is the steering mechanism of technological society? Will technology subside into the background and let us concentrate on living, just as a great photographer concentrates on imagery and not on his machinery?

The most important question we can ask today is the question that should have been asked in the 1980s but wasn't: Given our technological possibilities, what sort of world do you want your children to live in?

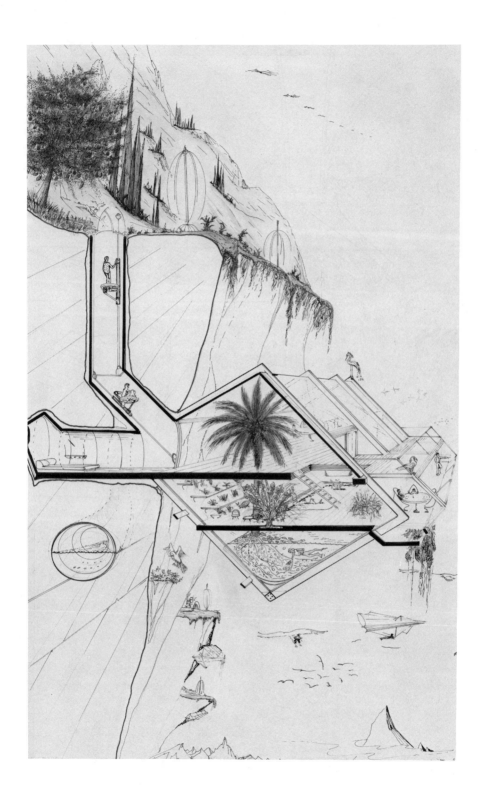

33

Installment Two

OUT OF WORK

PRESENT-DAY AUTHOR: One of today's greatest fears about automation is the fear of being put out of work.

Perhaps the biggest change in society in the next 30 years will be the change in jobs. Industries will not be competitive without robotized factories, full office automation, and corporatewide flow of data to decision-making computers. Many people today earn their living by doing work that can be done more cheaply by machines or work that will disappear completely as we move to more advanced forms of automation.

Some social commentators, alarmed by the prospects of robot factories and office automation, have stated that full automation should be prevented. However, a very probable scenario for the next 30 years is that the Japanese, who have already established a considerable lead in the use of robots and automation and who have already destroyed industries in the West by their aggressive marketing, will increase their dominance to such an extent that it becomes comparable to Britain's world dominance in 1840.

The West can hardly afford to slow down its acceptance of new technology. If it does, it will become an underdeveloped backwater, open to Japanese imperialism. No nation is now isolated. All nations are in competition with the most aggressive. This competition between nations has become a major power source heating up technology's crucible.

Seen from the viewpoint of A.D. 2020, work will look very different. A major question we should address today is, How do we get from here to there?

A key to success in the race for automation is the retraining of people to do new jobs. A person starting work today is likely to have to learn a new job not once but many times in his or her career. Some corporations, like IBM, have been highly successful in developing their workers and staff as automation advances. But in many cases the attempts at retraining have failed. A person over 40 who has been a clerk or production-line worker all his life is going to have difficulty learning a job that needs elaborate skills or intelligence. Attempts to train bank tellers to become programmers have often failed.

It is clear that we are moving to a society with a very different mix of jobs, *an intellect-intensive society.* There will be far fewer routine jobs and simple jobs, far more jobs demanding creativity and intelligence, far fewer jobs where work is drudgery, and far more jobs that people enjoy.

For this type of society to work well, *we have to develop human potential as rapidly as we develop technological potential.* This is one of the most important principles in our evolution to a more technological society. Human potential and technological potential must grow together. To grow technological potential without growing human potential will cause endless problems.

The future will be dramatically different from the past. As we invent the possible alternate futures, we must invent futures in which every person in society can play an important role. If we create a future in which half of society does not participate or is not employed meaningfully, this will be a society in which alienation and discontent will cause severe problems. If we create an automated society in which half the public is unemployed, the sociologists tell us this will be a society with great violence. It will not solve the problem to put the unemployed on welfare. People in society need to feel that they have a role to play.

Yet advanced automation is needed in both our offices and our factories; otherwise, we cannot compete internationally. Japanese goods will be cheaper and better and will dominate the marketplace. If goods from a country do not sell competitively, that country is doomed to severe economic problems and the unemployment that comes with them.

Many tasks that raise the quality of life cannot be automated. It would be nice to have much better theater, beautiful cities with parks and flower gardens, well-written television, magnificent music, symphony orchestras in every city, the sculpture of Michelangelo, the food of three-star restaurants in France, the landscaping of Capability Brown.

The problem with an intellect-intensive society is that not everybody has intellect. Not everybody can be engineers or computer programmers or can cope with the logic that technological jobs demand. We need to invent a future in which a high proportion of jobs are right-brain work as opposed to left-brain work—jobs that need the talents of artists, gardeners, entertainers, nurses, and chefs, jobs that need intuition, care, humor, love, and craftsmanship.

To have a high proportion of such jobs will raise the quality of life. Indeed, it is only after we have automated our factories and offices, eliminated the boring jobs, and killed off drudgery that we will be able to focus our full attention on the art of being truly civilized.

VIDEO: The same scene-changing sequence as in Installment 1 is used to slide from the present to A.D. 2019: a rushing sensation like flying very low at extreme speed through a mountainous valley, but the earth and hills look like endless etched microelectronic circuitry. Electronic music is used, evocative of time travel.

VIDEO: A car assembly line, with the work entirely performed by robots. Not a man in sight. Beautiful machinery. A sense of power and speed.

There is an explosion. The robots disintegrate. Car bodies are hurled everywhere. Chaos. The intricate machinery is reduced to charred and broken wreckage.

NARRATOR: (In front of a wall screen in her house showing the above image): In the 1980s, when the public began to realize the impact of industrial robots and other automation related to silicon chips, many people believed that this technology would put them out of work. In some places there were violent reactions to automation.

> *VIDEO*: A montage of jobs that can be automated: skilled workers, secretaries, printers, bookkeepers, mail carriers, assembly-line workers.

NARRATOR: Assembly-line work could be done better and cheaper by robots. Computers replaced middle management functions. Far fewer secretaries were needed because of office automation, especially when we had speech-input word processors and intelligent filing computers. Mail was delivered more cheaply by telecommunications. Printers lost their jobs as newspapers became set electronically.

> *VIDEO*: A visual image supports each function mentioned. Then we see welfare lines. A pair of vast hangar doors close slowly as despondent laid-off workers leave. Depressed faces. Anger. The hangar doors meet with a loud, very final-sounding clunk.

NARRATOR: When the public first began to comprehend silicon chips, they seemed like a miracle. Those chips were as crude as Watt's first steam engine. A fast evolution was to occur. The density of cells in an insect's brain was immensely greater than on the chips of the 1980s, and the insect's software was incomparably better.

> *VIDEO*: We see a column of the tiniest ants, several hundred of them lugging a piece of meat up a wall.

NARRATOR: Union leaders were highly vocal on the subject and made impassioned speeches about stopping or controlling automation. Here, for example, is Britain's Jack Jones, in the BBC Horizon film *Now the Chips Are Down* (1977).

> *Jack Jones*: A man gets his sense of identification from his job. When you ask him what he does, he does not say, "I'm a James Joyce reader" or "a Beatles fan"; he says, "I'm a welder, or a painter, or a bookkeeper."

> *VIDEO*: Shots of robots welding and painting; a terminal displays bookkeeping figures.

> *Jack Jones*: If we allow this to happen on a massive scale, it will not be the start of a great era, it will be the end of one. (Pause on the union leader's urgent face while the point sinks in.)

NARRATOR: How wrong they were. Automation of blue-collar and white-collar work happened on a massive scale, and it led to one of the greatest eras in human history.

It is not surprising that workers should band together to protect their jobs. This has happened throughout history.

VIDEO: Nineteenth-century Luddites, led by Ned Ludd, and their destruction of steam-driven spinning machines. Farm workers in Brazil rioting, burning machinery.

NARRATOR: The first steam engines were perceived by the workers as being a threat, just as were the first threshing machines and the first silicon chips. But these inventions led an era of unprecedented prosperity for countries that grasped the opportunities.

VIDEO: A montage of steam engines working, early industrial revolution industry, leading to the great wealth of Victorian times: magnificent homes, Covent Garden opera house. Victorian industry leading to the wealth of the 1920s: the Great Gatsby era, 1920s music. The microchip and robots leading to the great wealth of 2020. A thousand players performing Mahler's Eighth Symphony.

NARRATOR: Growth was as certain as sunrise. The old jobs were inevitably at risk. Some countries automated rapidly and remained prosperous. Other countries made a point of protecting out-of-date jobs and lost business *en masse* to the countries that did automate.

By the 1990s it was generally realized that it was not robots that cause unemployment but the failure to build robots.

Japanese goods flooded the West. They were cheap and of better quality. The Japanese automated faster.

VIDEO: A montage showing Japanese automation. Japanese cars. Headlines indicating 34% of the United Auto Workers' members out of work. A sequence of diverse goods saying "MADE IN JAPAN." A mass-production line of advanced robots in Japan. Western factories closing.

NARRATOR: Not content with automation at home, the countries with the most advanced robot production lines built factories abroad. The old factories could not compete. The profits from Japanese plants abroad went back to Japan.

VIDEO: A robotized IBM production line in Austin, Texas.

NARRATOR: This production line made computer terminals. In 1983 it produced 700 terminals per day and employed 130 workers. In

1984, after a complete redesign, it produced 2000 terminals per day of a more advanced design and employed only five workers. The number of machines produced per worker was 74 times as high. The new terminals were many times more reliable. This redesign for automation saved further money by drastically reducing inventory held in the factory.

In the 1980s such a factory was exceptional, but it was clear that similar changes were to sweep through industry everywhere. The world became flooded with goods that were both cheaper and much more intricate in their design, making elaborate use of microprocessors, memory, and, later, artificial intelligence techniques. Superlative hi-fi music systems were mass-produced and brought the dynamic range of the concert hall into people's homes. Television sets were mass-produced with wall screens that gave pictures almost as good as those in movie theaters. Intelligent lawn mowers, robot cookers, and a myriad of home devices were built. A new era of mass production deluged the world with sophisticated goods.

At that time, incredible though it may seem to us, the average person thought of technology as something that took the interest and creativity out of life. Intellectuals, forgetting the drudgery of primitive life styles, advocated "getting back to nature."

VIDEO: A historical sequence with narrative showing how work has evolved:

We see the peasant farmer: The life style, as photographed, looks lyrical—at one with nature. But we then see what backbreaking drudgery it was. We see the stunted bodies of men and women old before their time.

Then the industrial revolution—the drudgery of mine and mill. An early production line. The hard work, monotony, and dirt. The stunted lives of the workers.

The men who dreamed up ways of speeding production: from Arkwright and his spinning jenny to Henry Ford and the first automobile production line. Arkwright and Ford made workers more productive. But while they made work less strenuous physically, they made the drudgery more monotonous. We see a sequence of production lines, as car manufacturers became more and more sophisticated, 1920–1980. We see the faces of the brutalized workers, the noise and dirt of Detroit auto factories.

First Production Line Worker: You're nothing more than a machine. Except if you was a machine they'd be more careful with you. Machines are expensive. They give more attention to that machine and you *know* that.

Second Worker: The line's got to keep moving. Nothing must stop it. If it stops for 20 minutes, they lose $100,000. You can't go to the bathroom. If you burn yourself with the welder or get a cut, you can't leave.

First Worker: There's no way I could do that job and think about what I'm doing. It would be impossible for me. Your mind's got to be somewhere else.

Second Worker: The guy next to me slipped and cut his head one day. He was hurt bad. Blood all down his face. I stopped the line and ran to help him. The foreman immediately turned the line on again. That's the first thing they do.

VIDEO: An intercutting sequence of bored, pallid faces as workers perform mindless tasks on the assembly lines. We see women threading magnetic cores with copper wires. The dirt and monotony of a truck fabrication plant. Women working at a chocolate factory production line.

NARRATOR: When robots were first used, they simply replaced some of the workers on the production line. Later the whole production process was redesigned. Groups of workers would accompany a car through multiple robot operations. The workers needed more training, but the job became less dehumanizing. The job of tending the robots was interesting and needed skill. Many workers came to regard the robots almost with affection. They gave them names and treated them like pets.

Robot technology improved rapidly. Robots acquired vision units and other senses. They were programmed to take more intelligent action. They were linked by telecommunications to coordinating computers and operator consoles. Entire production lines were built to operate automatically.

VIDEO: The sound of robot machinery is heard. A night watchman opens the door of a robot factory where machines are working in total darkness. As he swings his flashlight, the small circle of its beam lights up different machines busily at work.

We see a factory in earth orbit, where robots are manufacturing products for the genetics engineering industry. No humans are present. A lyrical, magical, mysterious sequence, set to music: a machine dance.

The camera pulls back to show that the scene is on the narrator's wall screen. With it behind her, she talks.

NARRATOR: The industrial revolution increased Britain's prosperity enormously but degraded the lives of most workers. Similarly, twentieth-century production-line jobs were degrading: Workers in factories were treated like components of machines. Later, office workers were treated like components of machines. That was, in a way, as degrading as the early days of the industrial revolution.

> *VIDEO*: We see bank tellers and supermarket checkout clerks of 1980 operating their machines, their faces blank with boredom.

>> *Bank Teller*: They don't want you to do your own thing. They want everybody to be the same. The customers have to go to the first free teller, so I don't get to know a customer and build a relationship. The bank wants you to be like the machine you're working with. I switch it on, start up the screen, look through the teller window, and say "Hello, I'm a robot!"

>> *Airline Reservation Clerk*: I have no free will. I'm just part of that stupid computer. The computer is so expensive that everything is geared to it. It computes how long you take to deal with a reservation. You have to get the customer off the phone quickly and immediately take the next call. It monitors everything you do and how long you take. If the customer asks for a window seat, I often don't put it in. I can handle more calls and get a better grade that way.

NARRATOR: As we survey the history of technology, it is clear that a little automation is usually degrading to human beings. A lot of automation, however, can result in highly interesting jobs.

With more advanced automation, customers talk to the bank computers themselves or make their own airline reservations from their home screen.

> *VIDEO*: An automated teller machine in an airport. A person using a home videotext screen.

>> *Factory Worker*: (at a screen controlling several robots) I love this machine. It amazes me that they could make anything so beautiful. I can program it on the screen to do just about anything, and it always knows what preventive maintenance it needs. They keep on releasing new software and parts. They've got a huge catalog. I can hardly wait to get the newest release each month.

NARRATOR: Automation had another important effect. It greatly increased the quality of the goods produced. A human who is made to work like a machine becomes bored and careless. On a production line, such a worker ceases to care about quality.

> *First Production Worker*: You call the foreman's attention to a mistake and he'll ignore it. A bad piece of equipment? Forget

it. The line's got to keep moving. There's no way you can be proud of the work. They don't care if you do it good or bad. You've just got to keep the line moving.

There are specifications for number of welds. Nobody bothers with them. (He laughs.) You just do what you feel like and let it go. They have inspectors, but all of us know these things don't get corrected.

Second Worker: When I go in to buy a car or pretend to, I just look at the car and I can see all sorts of things wrong with it. Welds missing here. Bad alignment. The trim'll hide it. When we make a mistake, we say, "Don't worry, some yo-yo'll buy it."

First Worker: It depends how you feel. Mondays are the worst. You get a bit behind and you're bumping into the next worker. The line's got to keep moving, so you shortcut. If you didn't, there'd be a chain reaction all down the line.

Third Worker: We got drugs real bad. It's the boredom. They pop pills; they smoke. The ones with dark glasses on. They put 'em on when they're high. Some guys buy lots of candy. Something sweet. It'll hold 'em over 'til they get a shot.

NARRATOR: With advanced automation it would be unthinkable to use the human worker like a robot. Routine work is done by machines, freeing human beings to be themselves.

In many third-world countries we still have early automation—and people rebelling against it. As these countries grow to full automation, they release vast numbers of people for truly creative work.

Take the case of the coal miner as an example. It would be difficult to imagine a more degrading and dirty job than that of the early miners.

VIDEO: A miner in 1890 hacking at the coal face with a pick and shovel. Children of 12 working in the mines. We hear comments on mining from the writing of the time and see a montage of photographs.

NARRATOR: The early miners worked 12 hours a day. It was quite late in the twentieth century before the hacking of coal from the coal face was mechanized. The early coal cutters were noisy and created excessive dust. Manual picking of the coal and cramped mine tunnels were still needed. It was an inhuman job, yet few workers resisted automation more forcefully than coal miners.

NEWSREEL: The dirty, dehumanized faces of miners leaving a pit in Britain. Film footage of filthy work on the coal face.

Dissolve to a control room like a NASA space-shot control room. Thirty controllers sit in front of color screens at desks with keyboards. A controller flicks from a picture of a robot coal cutter to a picture of a robot loading coal onto a conveyor belt and then to a chart on which she adjusts a cursor with a mouse. She presses buttons. The robot cutter retreats and then attacks another area of the face. The coal on the conveyor enters a hopper.

NARRATOR: Most coal is converted into other products, such as coal gas, fuel bricks, plastics, and chemicals. As automation advanced, it became economical to process the coal at the mine location.

VIDEO: A controller looks at four picture charts on his screen with the divisions oscillating like instrument needles. A section of a chemical plant is seen. A controller makes an adjustment in his desk console. In the underground plant a servo valve controller hisses. Aboveground, all there is to be seen is a group of 20 pipes surrounded by cows in a field.

NARRATOR: Automation affected the job of the doctor in a very different way from that of the miner, but again it is true that a little automation tended to degrade the job of the general practitioner, whereas advanced automation greatly enhanced the job.

VIDEO: We see the beginnings of medicine: the doctor as faith healer, European prescientific practice, African tribal doctors. We see the beginnings of medicine treated as a science, where, if anything, the doctor is even more of a quack than before. We hear comments from nineteenth-century writing. We see and hear reactions on the huge growth in medical science.

NARRATOR: From 1950 on, medical equipment became more and more expensive and specialized. The diagnostic success rate of the general practitioner was low. More and more medical care migrated to specialists. The general practitioner felt increasingly demoted, yet the ongoing human relationship with his or her patients was important.

Primitive *artificial intelligence* began to be applied to medicine in the 1970s. At first progress was slow. The systems were inadequate. An AI system for mycin therapy (treating a complex bacterial ailment that affects mainly the stomach) was an early success and was responsible, in part, for the birth of expert systems technology. As the mycin system became perfected, it achieved a diagnostic success rate higher than doctors who were not mycin specialists.

A general practitioner needs to know about a thousand areas of medicine more or less complex than mycin therapy. There was no way that a physician could have even a fraction of this knowledge without access to computerized knowledge bases. By the year 2000 the doctor had a portable workstation that vastly expanded the ability to use medical knowledge.

VIDEO: Doctor and patient in A.D. 2019. The doctor has a black case in which to carry tools and medicines. In the lid is a computer screen that displays facts, images, decision trees, and an expert system dialogue. Medical instruments are connected by wires to the case.

NARRATOR: Given this computerized help, the general practitioner could concentrate on what a doctor was good at in the old days: the warm human empathy with the patient, the compassion, the bedside manner, and the human intuition that no computer can replace. The general practitioner in some ways has more in common with a nineteenth-century doctor than a late twentieth-century one. But unlike nineteenth-century doctors, modern ones have enormous computerized knowledge and inference power at their fingertips. Given the computerized help with the technical aspects of the illness and the patient's history, doctors can concentrate their main energies on understanding patients, their general health, their will to be well, and so on. Doctors can concentrate on being *healers*.

VIDEO: We see charts and graphs on the workstation screen and a doctor, like an old-fashioned healer, with a hand on the patient's head.

NARRATOR: With early automation, human craftsmanship was not respected much. As the tools became highly sophisticated, the value of the human touch was appreciated more.

VIDEO: The following sequence is illustrated with photographs.

Man in White Coat: My family have all been in furniture manufacturing. My great-grandfather was a carpenter. He made beautiful furniture. We still have some of it. In the old days a carpenter did the entire job. He designed and built the entire piece. He was a craftsman.

His son had to work in a factory. The development of more and more efficient production meant less and less involvement of the men on the production line with what they were making.

VIDEO: We see how in a widely assorted range of products, the method of production became more and more similar—more and more a depersonalized assembly line, less and less product-specialized skill required.

Man in White Coat: (with photographs) My father worked in a factory too, but by then it was all robots. At first they only churned out the same pieces over and over again. My father did various jobs including toolmaker. He used to think grandfather's job was inhuman.

When I was young, I was horrified at the idea of working in a factory, so like my great-grandfather, I have my own workshop. The public today wants quality and craftsmanship. They like antiques made before the twentieth century because of their craftsmanship. I use a dozen different robots. They polish and buff to a degree that would drive a man crazy. They cut and assemble with perfect precision. But the craftsmanship is back. I design the furniture, design intricate carvings on my computer screen, and see that it is built with the highest level of care and quality. The inlays and veneers of different woods are done by robots and are as elaborate as I want to make them. You can go to museums and you don't find pieces as beautiful as come out of this workshop. I love my job.

NARRATOR: (at home) By the 1980s Western society was in a mess over its attitude toward work. Many jobs were unsatisfactory in ways that could only be relieved by automation. The miner's job needed automation because it was filthy and unhealthy. The doctor's job needed automation to help manage knowledge too vast for the human brain. The fabricator's job needed automation because of the soul-destroying boredom of the production line.

Machines took over the drudgery first from men's muscles, then from men's minds. The first stages of automation usually degraded work; full automation freed men to be creative.

For a decade people had bought cheap chess-playing machines at department stores and accepted that almost nobody could beat the machine unless it was set to play benevolently at a low level. Few people asked: If a $50 machine can do this, what could a $50 million dollar machine do?

In one field after another, computer decisions became better than human decisions. First chess, then applications like crew scheduling in an airline, optimizing the movements of tanker fleets, and assisting doctors in specialized diagnosis.

VIDEO: Sequence showing how sailing ships returned to the oceans because they were more economical than oil-burning ships for slow cargoes. But the captain no longer decided the route. This was done by a computer on land that had information on worldwide wind patterns and currents. Even an old sea captain had to take orders from a computer.

We see a satellite antenna on the ship and the captain's equipment on the bridge.

NARRATOR: It dawned very slowly on the public and the press that machines were becoming capable of making decisions that humans could not challenge.

Factory managers in some multinational corporations used programs created at the corporate head office. They were free to disobey them if they wished, but if in doing so they achieved results below target, they were in trouble. So most obeyed the machine and became reliant on it for certain types of decisions.

The issue of machine superiority became public when a doctor was sued for malpractice for disobeying a computer's diagnosis and therapy recommendation.

The press began to ask, Who is really in charge? Are the machines taking over?

The popular wisdom about a Frankenstein's monster–like machine was that you could always pull its plug. We were in charge because we could switch the machine off. However, it became clear that in many cases we could not do this.

You cannot pull the plug on the telephone network or on defense systems or on life-support systems. Even low-level computer systems for administration became indispensable. There is chaos if they fail to work for long. Today millions of computers are interlinked, like the telephone network. We are totally dependent on this vitally important network and certainly could not switch it off.

Science fiction writers have claimed, sometimes in apparent seriousness, that our days as the dominant species on the planet are limited. Machines will take over from us.

VIDEO: Shots of Arthur Clarke saying on a 1980 *Nova* program about machine intelligence, "The Mind Machines," that the machines may keep us around as pets.

NARRATOR: The reality is more subtle. The computers do not "take over" but are increasingly difficult to disobey in business and professional situations. The servant remains a servant but, like Jeeves, has become indispensable and has acquired a talent much greater than that of the master. This is not necessarily alarming so long as the rules are clear and the computer debugging is thorough. It is disastrous if the rules change and the program does not.

Sometimes we tend to stop thinking and just obey the fossilized computer programs. But this is what bureaucrats have always done with their rule books. It is a delicate balance.

A machine can play chess, diagnose illnesses, or control an airline fleet, but these are relatively precise, specific functions. Today's machines can carry out a million such functions better than we can.

But they cannot correlate the different functions. In the foreseeable future, computers will be able to do almost any specific logical task better than humans, but they will not be able to integrate, synthesize, or be creative. The machines have access via data networks to a million types of software. But they cannot get their act together. Getting the act together seems to remain a function of biological brains, not electronic ones—at least for the time being.

The task of the human manager is to apply common sense. Common sense, we hope, will prevent misuse of highly complex systems and apply judgments about the reasonableness of their results. Human intelligence, then, is so completely different from machine intelligence that we should not use the same word for both. Both are vital in today's society and have to work together. Humans in many jobs had to learn how to work acceptably with machines that had immense logic capability.

NARRATOR: An area of work that had the greatest effect on our environment was the work of artists. As factory and office work was increasingly automated, more people entered the artistic professions.

It was once thought that artists were immune from the computer invasion. Nothing could have been more wrong. In the 1980s emerged the first personal computers that could display, manipulate, and print pictures. Crude software with which pictures could be created on the computer screen came into existence. This software improved and by the 1990s became impressive and popular. Optical disks could be purchased containing vast libraries of graphic images that could be manipulated, edited, and added to on the screen of a personal computer. Children, as soon as they could press a keyboard and work a mouse (which they do before they can talk), played in fascination manipulating the graphics on the screen.

> *VIDEO:* A one-year-old child happily playing with a complex image on a large color screen.

> Then the following narrative is illustrated visually.

NARRATOR: Expert systems for artists became one of the most fascinating areas of AI development. Used extensively by graphics designers, they gave new and at first startling imagery in advertisements, in package design, and on television. These expert systems dropped in cost fast because of mass-market popularity. A great diversity of expert systems and graphic-image disks flooded the market. It is possible to create images of any size that could be *scrolled* across the computer screen. The computer automatically adjusted images for perspective if necessary.

Versions of architects' software became popular in the home and converted many people into amateur architects. People designed their dream home, experimented with architectural form, created cityscapes, and landscaped their own gardens.

At the same time, robot machines evolved that did woodcarving, sculpture, stonemasonry, textile and carpet weaving, and so on. These machines improved until they could do highly detailed work with great precision. They were tireless at buffing, polishing, and finishing their work. The robot machines accepted instructions on disks from personal computers, so that objects designed in the home with intricate computer assistance could then be manufactured. Homemakers decorated their environment with superbly elaborate home-designed rugs, Grecian marble friezes and columns, pottery, fabrics, and furniture.

NARRATOR: (over video) The visual style of a new age was born. Elaborate computer-designed sculpture and mosaics and stonecarvings decorated new buildings everywhere. Homes and workplaces were draped with voluptuous textiles. Design with far more complexity than in earlier eras became high fashion. It was the age of the ornate.

The era from 1920 to 1990 became known as the Age of Blandness. It was the only age in history when such a massive amount of building had taken place almost completely devoid of decoration. The undecorated rectangular-box buildings became regarded as boring. To some they were the mark of an uncivilized era when society's activities were dominated by accountants, efficiency experts, and the need to maximize productivity.

Building sculpture, magnificent paneling, resplendent tracery, and intricate hangings were turned out in great quantity and infinite diversity by the computer-controlled machines. Elaborate chandeliers, twisting suspended staircases, delicately carved banisters, and rich tapestries with computer-generated fractals were the hallmarks of a new era.

Female fashion became equally decorative. Exquisite necklaces and dresses with rich patterns of tiny sequins were made by the robots working through the night with infinite patience.

NARRATOR: When automation and microchips first began to spread, governments, unions, and television programs created scary visions of societies with mass unemployment.

The reality was different. A vast amount of work was needed to create the new society. The main factor holding back the most valuable uses of new technology was shortage of people. Often, however, it was a new type of work for the people involved. So the great problem was not long-term unemployment but retraining. Vast numbers of people needed to be retrained to perform different types of jobs.

Some people were able to shift their type of employment relatively easily. Executives, professionals, and intellectually skilled people were in great demand and could usually adapt to new areas. Unskilled workers who were not highly paid moved to different categories of work, and when they learned to do more skilled functions, such as looking after complex machines, their pay increased. The category of worker most hurt by the rapid changes was the skilled worker who did not want to change his or her skill. Typesetters and printers, for example, often took great pride in their craft, then suddenly found the craft no longer needed because of printing automation.

VIDEO: Printers at work setting type and composing pages, dissolving to shots of word processors and editors at work on a screen.

NARRATOR: Skilled workers either had to learn a new skill quickly or find themselves unwanted or unable to earn the same pay. A 40-year-old printer with a large mortgage to pay was in trouble unless his management or union looked after him.

Most societies changed too fast for social comfort. Those that changed more slowly lost out to foreign competition, and then their employment problems often became worse. The rapid rate of change needed compassion, good management, and substantial planning. Often the most dangerous thing a society could do was to bolster dying industries or dying forms of work artificially.

VIDEO: A montage showing the new forms of work.

NARRATOR: The new work needs were immense: building the new types of factories, redesigning all products to take advantage of microelectronics, converting the world's textbooks to computerized and videodisk courses, developing the computer-aided medical diagnosis systems, cleaning up the landscape, rebuilding gardens so that machines could tend them, creating new energy sources and new food sources to keep up with the world's burgeoning population, making the waste and desert areas of the world productive, creating the food products, sauces, and software for the household's automated cooker, and creating the fabrics, mosaics, architecture, clothes, ballrooms, sculpture, and embellishments of the age of robot-built elegance.

NARRATOR: It was sometimes thought that people would not want to work in an age of robots.

NEWSREEL: Interview, 1985.

Interviewer: Suppose factories were run entirely by robots and people were free to do what they most wanted.

Factory Worker: Maybe they'd like to have a shop or farmstead— but land's running out. Maybe they'll just sit by a river and fish. I'd say it'd be a total disaster. People'd just give up.

It wouldn't be safe on the streets. You'd have loafers, people on dope, bikers. The streets would be full of people with unoccupied minds, and unoccupied minds get occupied with trouble.

NARRATOR: There have always been slobs in society. We see them in Brueghel paintings, read about them in Victorian novels.

VIDEO: Drunkards in paintings by Brueghel and the Dutch masters. Scenes of Bowery bums.

NARRATOR: With the reshuffling of work caused by automation, many people could have dropped out. They could have sat at home with a case of beer and watched television. In fact, the ratio of dropouts has not changed much. It is about the same as in Shakespeare's time, in Victorian England, and in the 1970s. Human nature has not changed since Shakespeare's time. The same human nature is set in a shatteringly different technical and social environment.

Most people, on the whole, like to be kept busy. What was wrong with the so-called Protestant work ethic of the nineteenth century, in particular, was that it regarded being busy as in itself a virtue, not considering what the busyness was about. We can now see that it is creativity, new ideas, and new methods of working that increase the quality of life and prosperity, not grinding away in the same routine. And creativity and inventiveness need time.

Instead of the streets being full of unoccupied minds, there are now more occupied minds than before automation.

NARRATOR: Certain aspects of work become automated in an age of intelligent machines and computer networks:

VIDEO: Narrator reads list as it rolls upward.

- SIMPLE TASKS
- ROUTINE TASKS } (both physical and mental)
- REPRODUCIBLE TASKS
- TASKS REQUIRING CALCULATION
- TASKS REQUIRING COMPLEX LOGIC
- TASKS REQUIRING VAST KNOWLEDGE
- EXPERTISE THAT CAN BE REDUCED TO RULES

NARRATOR: As most such tasks became automated, automation played havoc with society. The rate of change was fierce. Other aspects of work, however, were not automated:

VIDEO: Narrator reads list as it rolls upward:

- ORIGINALITY
- INTUITION
- INSPIRATION
- ART
- LEADERSHIP
- SALESMANSHIP
- HUMOR
- LOVE
- FRIENDSHIP

NARRATOR: Many jobs encompass aspects from both of these lists, so they are partially automated, making the individual more powerful or productive.

It is clear that as we increase technological potential, it is also vital to increase human potential. Humans must be trained to be more creative, more skilled, more compassionate, more caring, more civilized. We must think about the civilization we are building. It is surprising how little thought was given to that at the outset of the robot revolution.

Robot systems automated some nursing tasks in hospitals.

VIDEO: Film of robot nursing stations.

NARRATOR: In World War II, hospitals near the battlefields dealing with wounded soldiers came under great pressure. Makeshift hospitals were set up in which all of the nurses were men.

NEWSREEL: Pictures of male nurses tending soldiers.

NARRATOR: The hospital records show that most of the soldiers tended by male nurses took much longer to heal than those tended by female nurses. Often their wounds took three times as long to heal. It is clear that the attentions of an attractive female made the soldiers' wounds heal faster. No automation can replace the effect of the nurse who cares.

The same is true with teachers, actors, sales representatives, executives, and entrepreneurs. The machines can make them more efficient but cannot replace the drive, love, originality, aesthetics, or leadership of the human being. The *great* actor, or artist, or orator, or leader, has mental powers of *great* subtlety which artificial intelligence has no hope of imitating.

A great society needs synergy between the power of intelligent machines and the human qualities that cannot be automated. These human qualities grow and improve enormously with training and encouragement.

The effort to improve human potential must go hand in hand with the effort of improving technical potential. To do otherwise leads to disaster.

VIDEO: A montage of vigorous, creative, compassionate human faces at work.

NARRATOR: The rate of creative production in the arts and sciences is a hundred times what it was in the 1980s. The world population has doubled. Fifty years ago only about a billion people lived in societies affluent enough to encourage creativity, and only about one percent of them indulged in creative pursuits. Now 5 billion people—more than half the world's population—live in such societies, and about a quarter of them are creative. Children are trained for creativity. They grow up with intelligent machines that help them. When a person works at a computer screen to produce a creative design, the machines can mass-produce that design in vast quantities.

The rate of creative production is 100,000 times that of Athens at its prime or Renaissance Italy. The sense of excellence is nothing like as high as in Athens or Renaissance Italy, but we have the means to change that if it were the public will.

Telecommunications, laser sculptors, and optical disks spread the creativity worldwide at great speed. The best of one country's culture is quickly reproduced in every other country. Robots worldwide weave Chinese textiles, sculpt Italian designs, and make French stained glass. Events like the Carnival in Rio spread from city events to world events.

VIDEO: Frenzied scenes of samba dancing and parades with spectacular costumes and dance, like in Rio at Carnival, but now in Berlin, Perth, Vancouver, Paris. A bank of video screens showing carnivals around the world. Vast crowds who cannot stop sambaing. Scantily dressed women dancers, floats with massed flowers, outrageous costumes, men dressed in ostrich feathers. The Golden Gate Bridge with half its traffic lanes closed for samba festivities. Perspiring faces and bodies that cannot stop moving to the wild infectious music from Brazil.

NARRATOR: In more and more careers it became impossible to distinguish between work and play. Artists and entertainers had always lived like that. Now there is more scope for them. Through the twentieth century more and more scientific jobs became like that: creative scientist, inventor, zoologist working with wild animals. Athletes were next, in the middle of the twentieth century. Professional athletes became, in a way, entertainers. More and more sports generated sports superstars. Baseball was the first; then football, tennis, soccer, golf, track, basketball, swimming, and gymnastics.

Once drudgery is taken away and creativity encouraged, more and more jobs become fun.

VIDEO: A teacher, sitting with a colleague in a beautiful landscape, apparently picnicking. Then we see his briefcase, with a keyboard and a screen in the lid. We see that the two of them are working on a series of computer-assisted designs.

A montage that says that work is fun: the gardener directing his garden robots; the engineer building a bridge with computer-assisted designs; the robot-assisted chef cooking a gourmet meal.

NARRATOR: Today, a much more affluent era because of automation, work is perceived as something that should be a challenge in life. The robots and computers and new industries make this possible. One of the great joys in life is that of being a good craftsman, artist, or creator. Now that most of the drudgery has been removed from work, there is scope for most people to be creative. Children grow up with computers that challenge them to compose music, design buildings, create animated cartoons, and use a multitude of expert systems for writers and artists.

Creativity takes many forms, ranging from the flower gardener to the biophysicist. Everybody can be creative at something. The absence of drudgery gives time to concentrate on the uniquely human aspects of jobs or on work that requires care, affection, and love.

VIDEO: Shots of a flower gardener, a teacher teaching one child, an orchestra rehearsal.

NARRATOR: Freed from drudgery, a substantial portion of society indulges in endeavors that require much education. Education continues throughout life. It does not end at 22 or so, as used to be the case.

Every town has its own orchestra, theater, and television studio. Small-town television directors can employ the special-effects facilities of the large cities by telecommunications. Musical theater spread everywhere with many people going to tap-dance classes. The world's best budding actors tend to migrate to the cultural centers for training.

Today's cities have magnificent flower beds and parks. Automation took the backbreaking work out of gardening, leaving gardeners to do what they loved.

VIDEO: Along with the park scenes we see a truck labeled "Landscape Automation." Joggers run past the flower beds.

The camera pans across buildings with giant walls of intricate mosaics, like the Taj Mahal, then shows a robot fitting the titles in place. Dissolve to a person designing mosaics on a large color computer screen.

NARRATOR: Something in humanity responds to the challenge of the mountain and goes out to meet it. Now that machines do our work, the struggle is the struggle of life itself to create a better world. We no longer live to work and make money. We live for the adventure of life itself. What we get from this struggle for new achievement should be sheer joy.

Without the old-fashioned drudgery of work, we need to redefine the goals of life, the goals of society. We need to ask what we mean by civilization. What constitutes a great civilization in an age when machines are intelligent?

VIDEO: A montage of avocations, showing connections between individualists: the coal miner who is an expert on orchids and powered hang gliders; the maker of artificial limbs who uses robots for wood-carving; the free-fall parachutist and organizer of old people's exercise groups; the medieval historian who learned to create medieval gargoyles as good as the originals; the doctor who built an expert system knowledge base of herbal medicine collecting data from Tibetan monks; the psychiatrist who devoted his life to editing newsreel tapes showing what human events led to war and how it could have been prevented. **Magnificent, triumphal, choral music.**

Installment Three

THE LITTLE RED SCHOOLHOUSE

PRESENT-DAY AUTHOR: In an era of extreme technical and social change, few things are more important than good education. As technology advances, humans will do work that is much more challenging and creative. The future mix of jobs will need far more talent and more education than today. In many ways, education in the West is not living up to the needs. Japan is doing a better job than the West, and this will have a major effect on their relative strengths in the future.

The society we portray in this series needs improved education and training at all ages. The education of young children needs to give them a foundation that will make them resistant to future shock. They need a set of values, attitudes, and skills that will enable them to ride the crest of changing work patterns.

My 9-year-old daughter went to a summer school and spent most of her time cutting shapes out of paper bags and similar activities. There were no computers, no intelligent books, no challenges of any type, but there was a very self-satisfied teacher.

In many schools today, children are simply not being challenged as they should be. Today's schoolchild will be at the prime of life in 2020 when the world will be something like that described in this series. Teachers everywhere ought to have that vision. They should be educating people to succeed at work and leisure in 2020. Unfortunately, teachers are often the last people to have such vision.

Many children do not use computers at school or, if they do, only play games on them. Many teachers lack the skills to teach interesting things with computers. Children from wealthy families grow up with computers at home and often become highly skilled with them, preparing themselves for the world they will live in. But this causes a gap in capability between rich and poor children that will become very significant to many as they reach college age or working age.

Education will not cease when school ceases. In most careers it will be necessary to retrain constantly. If a person continues to learn throughout life, his or her brain remains capable of learning. If we stop learning, the mental muscles atrophy, and retraining will be more difficult later.

Some of the images in this series give the appearance of a middle-class, not blue-collar, family. A challenge of today's electronics is that by 2020 there should be no working-class people. The conditions that created the so-called working class during the past two centuries will have disappeared.

It may be difficult for the blue-collar worker today to recognize that the home shown in Installment 2 could be *his* home, just as today's coal miner would have difficulty recognizing himself as the man who sits in the control room shown directing mining robots.

As computers spread into every nook and cranny of society, as an intellect-intensive society develops, it is essential to grow the abilities of people to participate fully. It is easy to visualize a society like Huxley's *Brave New World* with limitless opportunities for the intelligent and flexible but drugs and 100-channel television for the uncreative masses. Sociologically, this would not work. It would lead to violence or oppression on an extreme scale.

Today most people are not fulfilling anything like their potential. In many cases this unfulfillment exists because of need to work eight hours a day at a job that leaves little energy for anything else. We simply do not know what will happen when everyone in society is given the chance to develop full creative powers. The technology we are now unlocking challenges us to find out.

A most critical principle of society today is that human potential must be developed as fast as technological potential. In many cases we are failing to do that. It is a failure that needs to be urgently corrected.

VIDEO: *The same scene-changing sequence as before is used to slide from the present to A.D. 2019: a rushing sensation like flying very low at extreme speed through a mountainous valley, but the earth and hills look like endless etched microelectronic circuitry. Electronic music is used, evocative of time travel.*

VIDEO: A city school. The camera pans across windows with most of the glass broken.

NARRATOR: As we look back now at society plunging into the vortex of the computer era, it is astonishing to observe the negligence of its school systems.

VIDEO: A gang fight. Breaking glass. A horde of rampaging youths. Football fans rip a train to pieces. This sequence is intercut with the emotional violence of a punk rock concert.

NARRATOR: In the latter third of the twentieth century, many schools in the Western world became unworkable.

In some inner cities, some schools were places for gangfights, drug peddling, and organized teenage crime. Teachers abandoned the attempt to keep order. Few teachers stayed more than a year or two in one school. Reading, writing, and arithmetic scores dropped year after year.

NEWSREEL: (from 1979) A headmaster in a cap and gown is giving a lecture:

> *Headmaster*: Intelligence tests became questioned because legal-aid lawyers charged that they were culturally biased. We are spending less money on textbooks than on vandalism. Ability

grouping has been referred to as undemocratic. Grading has been called an arbitrary system of rewards and punishments meted out by authoritarian teachers. The professional literature is full of jargon-filled justifications for all of this, with scholarly footnotes.

Meanwhile, we put caps and gowns on people who are functionally illiterate. [1]

NARRATOR: Even worse, perhaps, most schools failed to adapt to society's new needs. Most teachers were unable to cope with the intelligent children's demands to learn new skills.

We have commented that in an era of artificial intelligence and robots the following aspects of work will be automated:

VIDEO: Narrator reads list as it rolls upward:

- SIMPLE TASKS
- ROUTINE TASKS   } (both physical and mental)
- REPRODUCIBLE TASKS
- TASKS REQUIRING CALCULATION
- TASKS REQUIRING COMPLEX LOGIC
- TASKS REQUIRING VAST KNOWLEDGE
- EXPERTISE THAT CAN BE REDUCED TO RULES

NARRATOR: The mix of jobs in society changed at a furious rate. No era in history had demanded more retraining of the work force. No era in history needed more skills and inventiveness in young people.

VIDEO: Montage of new types of jobs: a hydroponic farmer, robot production lines, textile design at a computer screen, a NASA mission control room, a microscope shot of a needle piercing a complex cell.

NARRATOR: It had always been realized that classroom teaching was no good for certain skills—learning a musical instrument, for instance. Now it was realized that classroom teaching was often counterproductive for many subjects.

Fortunately, when technology creates new problems, it often provides solutions to those problems at the same time if only the solution can be grasped and put to use.

The first videodisk came out in 1979. Its use for showing films on television sets spread slowly. In its early days few people grasped the potential of the videodisk. The compact disk became a mass-market product for music and then a powerful interactive medium for education.

In the history of humankind there have been three great media. First the book, and in general the printed word. The book is still a great medium.

VIDEO: Narrator indicates her home library in the background.

NARRATOR: Second, film and television. A great medium that can show all manner of things that cannot be shown in print, all of which has had an increasingly important impact on our culture and knowledge.

Third, the computer. The terminals in every home allow us all to use its software and data bases often without even wondering how the results are obtained.

The importance of the interactive optical disks was that in effect they allowed these three fundamentally different media to be combined in one cheap machine. The person who created material could for the first time interweave text, television, diagrams, speech, and computing on one screen, with the viewer fully interacting on a keypad.

VIDEO: A classroom with each student using a screen and keypad. We watch one of them. The teacher reviews the subject with the students one at a time.

We see a home with a medical student examining a patient on a wall screen. The patient's responses are filmed on the interactive disk. This sequence of interaction continues for some time.

NARRATOR: By the late 1990s the capacity of the videodisk had grown to 80 billion bits of data. It will store two hours of high-resolution television, or 216,000 frames of information. Eighty billion bits is enough to store a vast encyclopedia with color pictures and sound sequences, or 10,000 full-length novels. Its production cost dropped to less than $1. The price paid in the shops was mostly royalties. Jukebox-like machines provided random access to hundreds or thousands of disks, and in big libraries to millions of them.

It took a decade to learn how to employ such a powerful medium, leading to an industry far larger than Hollywood at its peak. A half hour of instruction might cost $200,000 to make, but the sales were huge. The impact on education and training of all types was enormous.

The main constraining factor ever since has been shortage of people to build all of the courses and products that humanity needs. It is ironic to reflect that when microcomputers were first perceived by the public, there was widespread fear of mass unemployment.

Ever since, the new industries have been desperately short of trained people.

VIDEO: Videodisk training in use to teach persons who service robots how to deal with malfunctions.

Interactive videodisks being used in the third world, where teachers are in short supply.

NARRATOR: The information on the disk was inside tough plastic and was read by a laser. This made the disk virtually indestructible in normal use. It is amazing to reflect how susceptible to damage were phonograph records prior to the optical disk era.

Small, laser-read compact disks became popular in the late 1980s for hi-fi reproduction and then for interactive computer use.

VIDEO: A person at home exploring in detail the magnificent stained glass windows of the Sainte Chappelle chapel in Paris.

NARRATOR: Libraries grew up so that the public could borrow music disks, educational television, art disks, or computer software. Rather like milk delivery men in an earlier era, entrepreneurs established disk delivery services to the home, allowing people to borrow films, software, or video games of their choice.

VIDEO: An optical disk delivery service in a rural Indian village with primitive houses. Excited children.

NARRATOR: The world's great teachers were in enormous demand to create optical disk courses. The best of them earn more money than film stars. Some today have vast royalties still accruing from classic disks made 30 years ago.

On the other hand, the lecturer who delivered one-way lectures in front of a chalkboard was doomed. No chalkboard lecture could compare with the skillfully crafted disks unless there was heavy interaction with the students.

The purpose of live human teaching became to interact, and this was carried out best on a one-to-one basis, as it has been for centuries at Oxford and Cambridge tutorials.

VIDEO: We see a fast sequence of Oxford tutorials through the ages: thirteenth-century Latin, fifteenth-century theology, nineteenth-century chemistry with equipment, twentieth-century nuclear mathematics at a chalkboard, and twenty-first-century spacestation engineering at a wall-screen computer terminal.

NARRATOR: The videodisk alone would have revolutionized education. Combined with the computer, its impact was far greater.

The computer was to have an effect very different from that of the videodisk. Computers in education are often used to "program" the children. It is important to use computers differently so that the child programs the computer.

Young children could not use the early computer languages like BASIC. However, starting in the 1980s, languages and forms of screen interaction came into use that enabled *very* young children to give instructions to computers. These forms of child-computer interaction became much better and more powerful at a rapid rate.

> *VIDEO*: A montage of scenes of children from 3 to 8 making computers manipulate images on screens. Use of a mouse. LOGO programs. A mechanical toy controlled by the computer. Little fingers pressing the keys. Animated faces discussing how to achieve a given result.

NARRATOR: When a child gives commands to a computer, the child is in control. The child teaches the computer how to behave. The computer, its screen pictures, its radio controlled animals and robots, is fun to play with. It is one of the most amusing toys around. Many small babies find it endlessly fascinating.

> *VIDEO*: A baby who cannot yet talk making color pictures move on a screen by touching the screen. The baby gurgles with delight.

NARRATOR: Computers are complex, so a child can make them do elaborate things, such as making them draw intricate patterns.

> *VIDEO*: A 6-year-old child using LOGO to create an elaborate diagram.

NARRATOR: In teaching the computer how to think, the child is beginning to explore what thinking means. The child thinks about thinking. This is a different type of learning experience from normal classroom learning, and is powerful in human development.

LOGO was a computer language for children created in the 1970s by Seymour Papert, who was interested in finding better ways for children to learn. With LOGO, the child gives commands to a device called a turtle, telling it to move a pen. Initially, it was a mechanical toy turtle.

> *VIDEO*: Children controlling a toy turtle, making it draw lines.

NARRATOR: Later the trutle was a pointer on the screen of a personal computer.

VIDEO: Two children give commands to the turtle on the screen, using LOGO. One types

```
REPEAT 20
PETAL
LEFT/40
END
```

First Child: That'll make it draw lots of petals.

The petals are drawn on the screen, forming a flower.

Second Child: It's gone around too many times. Make it 10 instead of 20.

First Child: No, I've got a better idea. Divide 360 by 20.

She changes the program to

```
REPEAT 20
PETAL
LEFT 360/20
END
```

First Child: It worked. Now we need a stalk.

She adds to the program

```
REPEAT 20
PETAL
LEFT 360/20
END
DOWN 40
```

Second Child: Let's make a superprocedure called FLOWER.

First Child: No, let's put some colors in first.

Dissolve to the situation somewhat later when the screen has three flowers with red petals and white stalks.

NARRATOR: When a child learns to program, the process of learning is transformed. The child becomes active and self-directed. The child adjusts his or her knowledge to achieve a desired result and then can endlessly modify or enhance the result. This acquired capability is a source of power and is perceived as such by the child. It is essential learning for children growing up in a computerized world.

VIDEO: The child grins triumphantly at a computer as it draws a whole garden of vividly colored flowers.

NARRATOR: The great learning theorist Piaget distinguished between concrete thinking and formal thinking. Concrete thinking relates to abstract or symbolic ideas, and in most children this does not develop until about age 12. Seymour Papert, in his explorations of children programming with LOGO, found that children could do symbolic thinking at age 6 with a computer [2]. The symbolic thinking became tangible because the child was making the computer draw pictures or move a toy turtle.

Piaget emphasized that different types of learning experiences occurred at different ages. A child was not likely to learn geometry or to think with symbols under age 12 or so. Seymour Papert demonstrated that the sequence of learning can be changed with computers, symbolic learning coming at a much earlier age. This drastically changed the teaching of mathematics, physics, and other subjects. Their abstract ideas became tangible and fun to play with. The horror of mathematics disappeared. Children became active builders of their own intellectual structures.

Computers and their software changed enormously in the years following the early uses of LOGO. Programs were made easier to build using a mouse, pull-down menus, and graphics. Sophisticated programs could be built without typing in commands.

VIDEO: Children using a Macintosh VI.

NARRATOR: Most of mathematics could be done by a computer. The computer could integrate, differentiate, factor any polynomial, solve simultaneous equations, and so on. The drudgery vanished from mathematics, and on a computer screen logical processes were usually built a step at a time, with much trial and error.

When a computer was used to teach calculus, the subject became highly graphical. Curves could be made to change on the screen. The child watched and played with the effects of growth and acceleration. The teachers concentrated not anymore on solving equations or the drudgery of differentiating but on the *meaning* of the curves, what it meant when they changed, and how they could be used to solve problems.

VIDEO: A montage of children with teachers helping them at workstation screens showing colorful and elaborate mathematical graphics. A picture of the earth and a small drawing of a rocket moving. Teacher expands a screen window showing a velocity curve.

Teacher: Let's make the burn time 50 seconds. Now, what's the terminal velocity? 20,000? OK, that's enough for it to leave the earth.

The window showing the rocket and earth is shown along with the velocity curve.

NARRATOR: The computer provided a great diversity of micro-worlds for the child to explore.

VIDEO: We see a boy of 10 designing an unusual suspension bridge at a computer screen. The computer does the stressing calculations, and although it is a suspension bridge, its road is not straight. It snakes in a S-shaped curve to link roads that can only run parallel to the river on opposite sides of a steep gorge. The suspending wires are not parallel and fan out from anchor points in the cliffs.

NARRATOR: When learning to speak or learning about the environ-ment, a child uses a very different type of learning from classroom education. With computers it became clear that many types of learning could be made more natural, like learning to speak. The child explores a microworld in the computer, a world designed to enable the child to learn in a fashion that is easier, less abstract, and fun. The child builds an intuitive understanding of the microworld, leading to under-standing of the larger reality.

NARRATOR: (over video) Old-fashioned classroom learning often created a phobia about a subject. Many children hated mathematics because they had to spend many hours doing sums or equations. Approaching mathematics via a LOGO-like environment makes it natural and fun. Imagine if dancing were taught by making children learn and draw diagrams of dance steps for hours a week before they were allowed to dance physically. Imagine making them pass a "dance facts" exam. Many children would have a phobia about dancing as a subject, and many natural dancers would not develop their aptitude. The same was true of the old ways of teaching many subjects. The old teaching methods, and often the subject matter, related to what could be drilled and tested with chalkboard and paper techniques.

Doing sums or writing equations is not an imitation of something that is exciting and recognizable in adult life. Building a simulated bridge and flying a spaceship on a computer screen are imitations of the adult world that can make learning exciting.

VIDEO: Children using a computer to help them write poetry, laughing at the word combinations they devise. Children analyzing a screen display of an electrocardiogram reading. Children composing and play-ing music at a computer screen. Children exploring a political decision with a decision analysis tree.

NARRATOR: As computers became widespread, powerful, and cheap, an enormous diversity of microworlds for education became available. Software for education became a major part of the publishing industry.

However, there was a problem. Most teachers could not keep up with the exploding computer environment; in fact, most did not want to.

Among children themselves, many became furious devotees of the personal computer.

> *VIDEO*: The den of a 14-year-old with a computer with high-resolution graphics, a modern videodisk unit, a TV set, compact disks lying everywhere, and many shelves of software.

NARRATOR: It became clear to many parents that a vital component of education was now taking place outside the schools. Parents tried to become involved, but most needed help. Private educators and consultants came into business to counsel and help teach children to use the new tools. Clubs were established in many towns. Children with computers became adept at using computer networks, and networks for this purpose grew worldwide.

Many children became astonishingly capable at an early age. Parents and educators worried about overenthusiasm. Nothing, it seems, would induce some children to leave their computer. They would work with it all night if allowed to. The drug of computing spread to increasing numbers of young people, and the software became ever more powerful.

As computers spread so rapidly through society and simple routine jobs tended to disappear, an individual who was comfortable and skilled with computers was much better able to earn money and be creative. Children who grew up with computers were better off. Unfortunately, in the 1980s and 1990s only certain children grew up with computers. Some schools moved fast to use computers in effective ways, but most fell far behind. Children without good computing at school or at home were at a disadvantage later. In many cases it was the children from wealthy or intelligent families that grew up with computers.

> *VIDEO*: An expensive home in the suburbs. A group of children using a simulation of a chemical plant displayed on a small screen.

NARRATOR: The gap between children from rich homes, with the latest machinery and software, and children from poor homes going to less-well-equipped schools became greater and greater. It became perceived as a social problem that violated the important principle of equality of opportunity. Poor children were at a major disadvantage with this powerful form of education. Often the problem was blamed on the schools, which had been so reluctant to update, reequip, and change their curricula and, particularly, to retrain their teachers.

As with many problems in society, entrepreneurs looked for solutions. Private schools and evening classes specializing in computer teaching flourished. Children with no computer at home could find one down the street at the new learning centers, where there were dedicated enthusiasts to help them. Many 14-year-olds earned evening money as instructors at the learning centers.

Worldwide franchises for computerized learning grew at a furious rate. The McDonald's of computer learning was called the Little Red Schoolhouse. Every town contained its Little Red Schoolhouse building, and these buildings were connected worldwide by a massive computer network. Little Red Schoolhouse, Inc., concentrated on poorer families and children who could not have the latest equipment in their own den.

Topics that seemed difficult in physics were made tangible and easy to learn. Topics that seemed difficult in high school were converted into computer and compact disk microworlds with which 12-year-olds could interact. Operations research and decision-making techniques that even advanced business school students once found difficult were made easy with computer graphics and taught to 16-year-olds.

It became more and more evident that the learning of complex skills ought to start at a very early age. If you don't learn to ski until you are past 40, you are in trouble.

A pioneer of child learning was Suzuki, who in the 1970s started to teach children the violin at the age of 2 and who also brought into the open something that seems so basic that it is astonishing that the twentieth century could have ignored it: A child will not learn properly unless the parent learns and practices alongside.

After Suzuki, the discoveries flooded in: Children were taught to swim before they could walk; children were taught several languages. Ideas in science were introduced to 5-year-olds as a variety of games.

VIDEO: Montage of 2-year-old learning the violin, 9-month-old babies laughing as they learn to swim, and a parent helping a 3-year-old use MacPaint.

NARRATOR: Little Red Schoolhouse, Inc., extended its franchise to teach musical instruments, computer-aided textile design, expert systems for artists, poetry, and anything with potentially high sales that traditional schools were neglecting.

New computer techniques were invented that could be taught to young children. A young child learns at a furious rate. Children of 18 months have a vocabulary of a few words. Two years later they have a vocabulary of several thousand words. For a time they learn many new words a day—and not in schools.

VIDEO: Japanese children aged 6 writing *kanji*.

NARRATOR: It would be almost impossible for me at my age to learn the basic 3000 or so Japanese characters. Few Westerners, even those living in Japan, do so after age 20. But the early computer languages and commands *were* designed to be learned by people over 20. New and more powerful command languages that could be taught to young children were devised. The computer generation gap became greater, with many adults not being able to understand what their children were doing.

VIDEO: Montage of Little Red Schoolhouse signs and group sessions with children. Complex symbols on computer screens, unrecognizable today, being manipulated by children.

NARRATOR: Classical educators fumed at the Little Red Schoolhouses and the high-tech dens of children. They stressed the virtues of classical education. A furious debate rages still today about the right balance among traditional schools, computer clubs, franchised education, videodisks, and the role of the home.

VIDEO: A children's orchestra and choir in a high Gothic cathedral doing a magnificent performance of Vivaldi's Gloria.

NARRATOR: One of the worst failures of Western education in the twentieth century was that it ceased to create in children a value system for living. Values are not instinctive; they have to be taught. Failure to teach them results in extremely expensive problems. The faster the rate of social change, the faster the passage into the unknown world of the future and the greater the need for a firm value system in a society.

When anthropologists look at primitive tribes, they expect them to teach the moral and other values of their culture to their children and to have strong means, ritual and otherwise, of reinforcing this teaching. Indeed, if they were to cease to do this, we would regard them as being on the way to cultural suicide.

Amazingly, many Western countries in the twentieth century did indeed cease to teach the moral and other values of their culture in schools. Walter Lippmann, a great journalist, in 1940 addressed the annual meeting of the American Society for the Advancement of Science:

SIMULATED NEWSREEL:

> *Walter Lippman*: During the last forty or fifty years, those who are responsible for education have progressively removed from the curriculum the Western culture which produced the modern democratic state. The schools and colleges have, therefore, been sending out into the world men who no longer understand the creative principles of the society in which they must live. . . . Prevailing education is destined, if it continues, to destroy Western civilization and is in fact destroying it. [3]

NARRATOR: The training in values became desperately important in the late twentieth century because the explosion of technology represented by robots, computer networks, artificial intelligence, drugs, and genetic engineering pulled the rug out from most of society's established patterns. Youth were alone in an uncharted jungle. Value education gave principles of how to behave.

The American psychologist Abraham Maslow stated in 1963: "The ultimate disease of our time is valuelessness. . . . This state is more crucially dangerous than ever before in history" [4]. Thirty years later it was to become far more dangerous than Maslow could have imagined.

Oddly enough, the primary reason for not teaching basic values in American schools was *religion*. The U.S. Supreme Court declared in *Engel* v. *Vitale* in 1963 that state-sponsored public school prayer violates the U.S. Constitution. Separation of church and state is an important tenet of the Constitution. Children of different religions are intermixed in the same schools. Religious teaching had been the primary vehicle for moral education outside the home. When this was removed, and in general as the power of religious bodies declined somewhat, moral education suffered and for many children disappeared.

VIDEO: Moronically bestial faces of teenagers mindlessly smashing windows. Punk rock images. A child thief snatching an old lady's gold pendant, cutting its chain, and running.

NARRATOR: It became necessary to replace the moral teaching of religions with general teaching of ethical values.

A worldwide study of value systems and religions identified 16 basic values that are shared by the world's major religions and cultures:

VIDEO: Narrator reads list as it rolls upward.

- HONESTY
- KINDNESS
- GENEROSITY
- HELPFULNESS
- RIGHT TO BE AN INDIVIDUAL
- SOUND USE OF TIME
- SOUND USE OF TALENTS
- FREEDOM
- TOLERANCE
- JUSTICE
- COURAGE
- CONVICTION
- HONOR
- FREEDOM OF SPEECH
- GOOD CITIZENSHIP
- RIGHT OF EQUAL OPPORTUNITY

NARRATOR: This set of values became the basis for a character education program for schoolchildren from kindergarten through sixth grade, designed by the American Institute for Character Education. The program, first used in the 1970s, included an explicit teacher's manual, wall posters, student activity sheets, and evaluation forms [5].

By the 1980s it was clear that schools that introduced the character education program usually produced a major reduction in vandalism and increased discipline, attendance, student morale, and scholarship.

NEWSREEL: Shots of many broken windows, destruction, graffiti on school walls.

First Teacher: We had terrible vandalism, theft, break-ins. We were concerned about pupil relationships to one another. After three years of our character education program, most of these problems have disappeared.

Second Teacher: We've had a 70 percent reduction of discipline problems referred to the principal's office for action.

First Teacher: Children like to relate their experiences with others. The program has made them look at themselves as others see them. With a few exceptions, the improvement in behavior is tremendous.

NARRATOR: In spite of the evidence, many educators argued against character education in schools.

It was sometimes thought that poverty or inequality led to social violence. However, in countries with high taxation for redistribution of income, such as Sweden, the level of crime, vandalism, drug abuse, and divorce was extremely high. Japan before World War II, on the other hand, was very poor, with great class differences, yet crime and vandalism were very rare. Oriental education has always been strongly concerned with the cultivation of character. The moral teachings of Confucius were important in China for many centuries from 500 B.C. until the twentieth century A.D. Confucian ethics can be regarded as independent of religion.

In Japan character education has been strongly emphasized in education. The teacher was valued as a person of high character who influenced the moral development of students both by teaching and by example. Even in poor areas of Japan the crime rate was very low. Japan plunged very rapidly into the era of computers, robots, and gene splicing, and its strong character education seems to have helped the extreme social changes. They happen with less crime, divorce, absenteeism, and drug abuse than in the West.

VIDEO: Columns of smiling, uniformed, disciplined Japanese school-children, walking down a street.

NARRATOR: It became increasingly clear that the role of the home was of vital importance.

In the late twentieth century, technology seemed to have produced an alienated society—children not talking to their parents, children unable or unwilling to do jobs their parents did, and parents unable to understand their children. Alienation had been on the increase since the Second World War. Literature was full of references to it:

VIDEO: A clip from *Rebel Without a Cause*, in which James Dean played one of the most famous early alienated antiheroes.

NARRATOR: From the 1970s onward in the United States and western Europe, there were various outbreaks of terrorism—young people against the system, very often rich kids.

NEWSREEL: Brief clips of Patty Hearst, the Manson gang, the Baader-Meinhof gang. The violence builds; we see shoot-outs.

NARRATOR: How did this come about?

VIDEO: A seventeenth-century peasant family, living together, working together, eating together. We see how education was a process of daylong apprenticeship—son to father or elder brother, daughter to mother or elder sister.

NARRATOR: After the industrial revolution, most families no longer worked together and ate together. At first the father, then both father and mother, went out to work. Children were compelled by law to go to school. The need to find work forced the family to move, following the breadwinner's search for a job. Home units became smaller, and grandparents were left behind or shoved into homes when they weakened with age and neglect.

VIDEO: Industrial revolution scenes. Victorian English governesses. Ships departing. Dissolve to Boeing 747 landing.

NARRATOR: When the jet age arrived, the rich as well as the poor had their family life disrupted. Business executives were forced to globe-trot, living for weeks at a time away from home. Children left home at earlier and earlier ages.

VIDEO: Rush-hour crowds; an office worker gulping coffee and grabbing a briefcase as he rushes from home to catch his crowded train. Dissolve to a small child being tucked into bed.

NARRATOR: The father would arrive home at 7 o'clock exhausted. He had little time to see his children. In many households he just settled in front of the television with a large drink.

VIDEO: The camera pans along ticky-tacky houses in a 1980s suburb.

NARRATOR: There are almost no jobs in this town. Everyone caught the train or drove into the nearby city. There was no theater, no museum or art gallery, no college. Bright children left home to seek education, culture, and opportunities elsewhere. Children did not return to live with their parents after college. Parents did not live with grandparents. The old were alone and neglected. The unified family of earlier times had broken up.

VIDEO: Picture of a Victorian family home and portrait.

NARRATOR: The soulless suburbs steadily sprawled out around large cities, their growth pattern very similar to the growth of cancerous cells in the body. Photographed from the air, the similarity of the pattern of suburban growth to cancer cell growth is striking.

VIDEO: Pictures comparing cancerous cell growth patterns with suburban sprawl patterns.

NARRATOR: No wonder the sense of alienation increased.

VIDEO: A montage to depict the dehumanized crowds: rush-hour workers, waiting lines for football matches or rock concerts, bored children in huge inner-city schools. The montage leads to a violent climax of street fighting.

NARRATOR: Violence grew more sophisticated as the rebels learned more technology and as more gadgets became available.

NEWSREEL: The development of bombs as used by Arab terrorists between 1985 and 1995. A hand pulls the lever of a small radio control unit. A boat explodes in flames. Headlines about assassinations of political figures.

We see the American terrorist gang that disrupted New York in 1998 with 12 bombs dropped down carefully selected manholes. All police calls, most Wall Street operations, and most of the electrical supply are stopped. Violent rioting and looting followed.

NARRATOR: The cost and discomfort of travel climbed and climbed for 30 years. At the same time, the cost of telecommunications plunged.

VIDEO: Three-foot satellite dishes in parking lots and on roofs. Sometimes there are three dishes pointing in slightly different directions.

NARRATOR: With advanced telecommunications there was little need to commute into the big cities. After all, what did people do when they got there?

VIDEO: Shot of New York rush-hour crowds. Montage of people telephoning, typing, having a meeting, using computer terminals, word processing. A clock says 5:30. More rush-hour crowds.

NARRATOR: They typed; they used machines; they telephoned; they had meetings—nothing that could not be done in their local community if it were wired with high-bandwidth channels.

VIDEO: Shot of hordes of bicycles pouring out of a parking lot past two satellite dishes. Children greet a father as he bicycles into his driveway.

NARRATOR: As electronics improved, more people began to work at home. Houses became equipped with spouse-proof, child-proof, sound-proof, tax-deductible offices. Children, as in earlier centuries, could see what their parents did for a living.

VIDEO: A child watching a teleconference meeting. A child watching a screen showing an architect at work using the computerized drawing tool.

NARRATOR: However, most people wanted company at work and preferred to work in a nearby office rather than at home. The local communities changed. They needed restaurants, sophisticated shops, and facilities for the local work force.

With amazing shortsightedness, expensive suburbs were built in the 1970s and 1980s without walkways. They were designed for cars to be the sole means of transportation. If you walked between buildings, you had precariously to dodge the traffic.

VIDEO: Shots of Oak Brook, Illinois, a smart new suburb of the 1980s with no pedestrian walkways. A jaywalker races across the street in front of speeding traffic. Sound of car horns.

NARRATOR: The passion for jogging in the 1980s made many people want changes in the design of pathways. Exercise should not be confined to the few people who jog. Walking became popular. People began to realize the pleasure of going for a walk and wanted cities and suburbs that were pleasant for walking. New urban areas had more parks. Walkways and traffic roads were separated. The clean cars of the twenty-first century did not cause pollution in tunnels. In new cities roadways were covered with earth and parks. In old cities areas with pleasing architecture were made into traffic-free precincts.

VIDEO: Ancient areas of Oxford and Salzburg with many pedestrians and no cars. Elaborate traffic-free shopping areas in Zermatt and Vancouver. Parks over roadways like the United Nations gardens in New York.

NARRATOR: Now that the energy and talents of the suburbs were not sucked away by the pull of the big city, community connections that had been missing before were built. People walked or bicycled to work and wandered into one another's homes. Pubs flourished. The community spirit and connectedness of small villages became the lifestyle of people. There were local theaters, skating rinks, public videodisk libraries, street musicians, cafés. A stroll through park-like shopping areas became a normal evening pastime. What had been shantytowns in remote but beautiful areas now became fashionable places to live, and much money was spent on restoring and improving them.

> *VIDEO*: Montage of bustling traffic-free streets full of alive human faces, street flirtations, an open-air café dance, antique shops, street singers entertaining, circles of people who gather. Satellite antennas in a Vermont village. Animated faces of children outside a Little Red Schoolhouse.

NARRATOR: Life is short, and commuting for two hours a day, 500 hours a year, was a mindless way to waste it. The growth of communities that combined work and living gave parents more time to spend with their children. The new forms of education needed parents' participation. Parents and children worked, learned, and played together more than in the nerve-fraying era of the commuter.

For most people the most important thing that they do in their lives is bringing up their children. To us it is obvious that parents need training and also that everything must be done to keep families together in an activity where work, play, and learning are practically indistinguishable.

> *VIDEO*: A father and son with a woodcarving robot. They are learning from a videodisk machine and can select a diversity of designs, which they can assemble into the robot's program.

NARRATOR: Parents were encouraged to spend two hours each day educating their children. They were encouraged to let the child see what they did for a living. The result was that collaboration between old and young became once again normal. The family became unified as it had not been since the industrial revolution.

Particularly vital, parents were encouraged to stress the importance of their children's value system.

Visual media became increasingly powerful as television screens grew to wall size, videodisks became cheap, cables and satellites provided large numbers of channels, and films became more graphic. Some children spent more hours per week watching media screens than they did in classrooms and churches. The media were a powerful factor in the character development of children and lessened parents' ability to influence that development.

People tended to emulate the characters on television.

VIDEO: Scenes from *The Deer Hunter*.

NARRATOR: The movie *The Deer Hunter* showed graphic scenes of people being killed playing Russian roulette in Vietnam. It was shown on American television in 1982. Twenty-nine people shot themselves in a similar way after viewing the movie [6].

The value system represented on television by the most popular films and serials was entirely different from the value system recommended for teaching to children. In particular, it idealized violence as being good.

VIDEO: Scenes from TV of brutality, machine guns, rape, strangling.

NARRATOR: Children learn to associate violence not only with hate and hostility but with virtue and justice. The role model for "good" behavior is often a man of extreme violence. The reasoning of this role model is usually only a gimmick. His method is to use violence. By using violence he succeeds. Children conclude that because they see it so often, it is probably an appropriate way to behave.

Numerous studies have shown that such television results in a value system that idealizes violence.

Frederic Wertham, a consulting psychiatrist at a New York hospital, described the television of the 1970s as a school for violence.

SIMULATED NEWSREEL:

> *Frederic Wertham*: If anybody had said a generation ago that a school to teach the art and uses of violence would be established, no one would have believed him. He would have been told that those who mandate the mental welfare of children would prevent it. Yet this education for violence is precisely what has happened. We teach violence to our young people to an extent that has never been known before in history. [7]

NARRATOR: The violence portrayed was to become much worse than when Dr. Wertham wrote those words. Pornography with extreme violence became common. As each new outrage became familiar with the public, producers tried to outdo it with something even more extreme.

Finally public protests led to action, but only after incalculable damage to two generations.

The difficulty lies in the abhorrence of an advanced civilization for censorship. Censorship might lead to dogma or to limitations set by people of questionable wisdom or taste. Athens in its prime had no censorship; nothing was taboo. Sweet reason pervaded Athenian society so that any subject could be discussed or written about rationally. However, while mature adults might be free to make their own decisions about whether they watch violence and pornography, the situation with children is entirely different.

In some countries technology itself brought the answer. All television was broadcast scrambled and could not be displayed without unscrambling. Parents and other owners of television sets decided which programs would be unlocked for viewing.

VIDEO: A user turns a key on the television set. Certain programs on a listing change from red to green.

NARRATOR: The TV listings came in videotext form. A person looked up what was to be broadcast and could unlock the programs selected. Programs could be unlocked several weeks ahead. If it was a pay TV channel, the subscriber was billed for unlocking the program.

With such systems, children can only watch programs their parents unlock, so the parent is responsible for the child's viewing. Some children, with the diabolical ingenuity of the age of computer hackers, manage to view programs they should not, but at least they know that the program is disapproved of, so they do not accept, say, a violent hero as an approved role model.

In some countries the commercial television authorities fought furiously to prevent program locking. Parents could buy computer-controlled sets for any channels and so were ultimately responsible. In many countries legislation was passed to give all parents television guidance and discretion and locking facilities.

VIDEO: Scenes in the narrator's city, which was built more or less from scratch on the north Australian coast in reclaimed desert, where technology was needed to make the sea safe for humans.

We see the semi-underground houses, with gardens on the rooftops. We see the abundance of flowers and fruit. Everyone possesses a garden of tropical abundance. The city looks like a park—its buildings are hardly visible. There is sculpture everywhere, mosaics, intricate textiles, and rich, complex designs on buildings. Factories are all underground, and there are very few people in them.

A river with punts and willow trees. Pubs by the water. People in animated conversation in the gardens. A family by the water. A family is building themselves a boat. The son and daughter are both working together on it, watching a video demonstration on how to assemble parts from a kit. They have a video machine outside by the boat, which they stop and play again as they require.

NARRATOR: No vandalism, little crime, children learning with their parents, no commuting, no pollution, but access to the world by telecommunications. Much work is done by remote control.

> *VIDEO:* An advanced dance lesson with one of the world's stars, the students in small groups in various studios around the world, the teacher giving instruction from in front of a battery of screens where she can zoom into close-up on any student as soon as she notices a faulty movement. We see a small child designing space-station structures with long cables between the components, switching up pictures of various structures of the past with all their stress forces marked. The child runs through possibilities at a great rate and decides on something very beautiful and strange. Magnificent choral music wells up throughout.

NARRATOR: A sense of all the world's knowledge and culture at our fingertips. A sense of time for your own family, your own immediate neighborhood. A sense of worldwide interconnection, both professional and friendly.

A sense of challenge. The possibilities of the new culture are endless. The children know it; their parents know it. Lessons in school, in clubs, in the home, in the privacy of a child's den.

A sense of values. Children taught responsibility. Agreement on values by schools and the media.

A sense of the power of the human mind and how in earlier eras it was not challenged, drawn out, strengthened, or exercised.

REFERENCES

1. Condensed from James C. Enochs, *The Restoration of Standards: The Modesto Plan*. Bloomington, Ind.: Phi Delta Kappa Educational Foundation, 1979.
2. Seymour Papert, *Mindstorms*. New York: Basic Books, 1980.
3. Walter Lippmann, *Education Versus Western Civilization*, address given at the annual meeting of the American Association for the Advancement of Science, December 19, 1940.
4. Abraham H. Maslow, "Fusions of Facts and Values," *American Journal of Psychoanalysis* (1963).
5. Frank G. Goble and B. David Brooks, *The Case for Character Education*. Ottawa, Ill.: Green Hill Publishers, 1983.
6. "What Is T.V. Doing to America?" *U.S. News and World Report*, August 2, 1982, p. 27.
7. Frederic Wertham, "School for Violence: Mayhem in the Mass Media," in *Where Do You Draw the Line?* ed. Victor B. Cline. Provo, Utah: Brigham Young University Press, 1974.

Installment Four

BIG BROTHER

PRESENT-DAY AUTHOR: In 1949 George Orwell wrote a book that made a generation shudder at the possibilities of totalitarianism: *Nineteen Eighty-four*. Its slogan was "Big Brother is watching you," and Big Brother, the government, knew everything.

George Orwell had never heard of computers when he wrote *Nineteen Eighty-four*. With computers the surveillance of human activity could be automated. There seems little doubt that past dictators would have used computers, surveillance, and data bases relentlessly if they had been available.

As we hurtle into a society where microchips are as common as nuts and bolts, can we retain our freedom, our right to be let alone? The technology that is emerging so rapidly and inexorably can be used to oppress us or free us. The public needs to know the dangers and opportunities for betterment.

It is dangerous to allow government, any government, to become too big and powerful. Large government bodies tend to breed their own worlds with their own rules, behavior patterns, and power structures, largely unrelated to the needs of the governed. Computer consulting work gives one a license to see the internals of many organizations. In private industry there are clear goals and critical success factors. There is a relentless pressure to make a profit. In government bodies it is often difficult to identify critical success factors. Making a profit is usually not a goal. Internal politics takes over and often creates byzantine goals of its own. Unless monitored very carefully, the empire builders justify the use of more and more people. They create regulations, forms to fill in, and procedures, which they justify vehemently but which in the broader view damage rather than improve the ability to make society pleasanter to live in. Nobody except the bureaucrats wants bureaucracy for its own sake.

Perhaps the worst danger of computers is that, used badly, they can be a relentless amplifier of bureaucracy. Parkinson stated that work tends to grow to fill the time available. Similarly, complexity tends to grow to fill the computer power available.

There is an extreme contrast between the image of a future society with electronically aided totalitarianism and a future society that uses technology to enhance the democratic process and create an Athens-like capability to debate societal issues. Both are now possible. By A.D. 2020 it seems likely that some countries will have electronic oppression and some will have electronically enhanced democracy.

The public must be fully aware of the dangers of overgovernment and must exert relentless pressure to avoid bureaucratic harassment and expense. This requires laws and a constitution that reflect both the dangers and the opportunities of advanced technology. Technology, used better than today, can greatly enhance a democratic process.

It is vitally important to be asking, What actions can we make *today* to maximize the probability of freedom *tomorrow*?

VIDEO: The same scene-changing sequence as before is used to slide from the present to A.D. 2019: a rushing sensation like flying very low at extreme speed through a mountainous valley, but the earth and hills look like endless etched microelectronic circuitry. Electronic music is used, evocative of time travel.

VIDEO: A car at night speeding along a highway takes a bend too fast and keeps going. We see a large truck coming in the opposite direction. Suspense. The audience knows there is going to be a crash.

Brazilian samba music is playing in the car. The windshield wipers push heavy rain off the glass. Through the rain and wipers we see the headlights swinging around rock cliffs.

Suddenly the lights of the truck appear head on. Violent evasive action.

The car plunges off the road, spins, and hits the rocks. There is a sudden cut in the sound track.

Sound of wind. Then an ambulance.

A medic is attending the driver inside the ambulance.

NARRATOR: The driver lived, because of telecommunications. The ambulance found him fast. His identification number was transmitted to the hospital.

VIDEO: We see a card like a bank card being put into a transmitter in the ambulance. Shots of antennas. A computer room.

NARRATOR: His medical records were stored a thousand miles away. By the time he arrived, the hospital was waiting. The staff knew his blood group and his allergic reaction to certain antibiotics.

VIDEO: Instructions to the ambulance staff spatter out on a small printer in the ambulance. They keep the patient alive and transmit electrocardiogram readings.

NARRATOR: A few weeks later the victim is fully recovered.

Clearly, medical uses of computers are of great value. It is comforting to know that the computers will sense if our car crashes and will have access to our medical records.

But the same technology can also store other information about us. Our credit records, the exams we failed, what videodisks we borrow, the numbers we telephone, how we use our terminals, our misdemeanors or suspected misdemeanors, our mail now that it is all electronic.

VIDEO: A goldfish in a bowl.

NARRATOR: Our lives could be almost completely visible to the computers. We could have as little privacy as a goldfish in a bowl.

VIDEO: Scenes from the film *1984* including the posters saying "Big Brother Is Watching You."

NARRATOR: A computerized society needs laws and controls to protect privacy. These laws started to come into existence in the 1970s.

In totalitarian countries we can now observe the appalling effects of computers used by the government.

In the 1990s more and more of the Russian empire became more and more angry at Russian rule. The Russians countered growing unrest with cyberveillance.

A centralized data bank for the entire empire was created with complete dossiers on everyone under Russian rule. These dossiers were available instantly in their entirety to the government and the police.

In exchange for the many privileges that party membership brought, every party member had to carry a radio paging device. This enabled him to be contacted at any time except when he switched off the receiver. Such devices have been used by businessmen in most countries for decades and were very useful. In Russia, however, the device also *transmitted* periodic bursts of radio lasting less than a millisecond. These were passed onward over data networks from relay stations located so that few areas escaped surveillance. The owner of the device could switch off the receiver but not the transmitter.

VIDEO: Polar bears roving the arctic wastes.

NARRATOR: As early as the 1970s these bears had tiny transmitters attached to them. They transmitted periodic bursts of radio that were relayed by satellite. Scientists could track all movements of the bears.

Similar devices were used in human medicine for constantly monitoring patients with weak hearts, epilepsy, or other ailments. The benefits and curses of technology are linked.

Any irregular movements of the party member could be detected by computer and checked. The computers noted when the paging receiver was switched off.

Every car also transmits its license number continually. Its whereabouts are known. This enables the owner to be automatically billed for parking in cities, billed for tolls, or automatically warned about speeding.

Many computer terminals, especially portable ones, also transmitted an identification number. The same pocket radio system had many uses.

The Russians were slow in introducing the electronic fund transfer systems used by banks in the West. When they did introduce them, they created a largely cashless society. This was efficient and saved vast amounts of clerical work, but it made possible the monitoring of all a citizen's financial transactions as desired. Any unusual transactions could be detected automatically by the computers. An underground economy based on barter grew rapidly.

Computerized telephone exchanges give many useful services. In Russia they provided a service not widely employed in the West. They filed a record of all numbers called and who called them. The system could note anything suspicious, such as phoning a known suspect. As soon as a person was suspected of subversive activity, all his phone calls were automatically recorded. Numbers from public call boxes could be dialed only with electronic identification cards.

Computer circuits were designed to recognize human speech. Such circuits had many applications.

> *VIDEO*: A person in a wheelchair instructing chair movements by speech.

> An executive says, "Hey, computer!" A digitized voice responds, "Yes, sir!" The executive says, "Display the sales figures for Hong Kong." Computer: "Monthly figures, sir?" "Yes." The figures roll down a wall screen.

NARRATOR: This speech analysis proved especially valuable in telephone or microphone surveillance. The microcomputers can be programmed to listen ceaselessly for key words.

Large libraries of storage were needed for recording all the surveillance data. The optical disk and optical tape technologies provided ideal media.

> *VIDEO*: A robot access mechanism races down an optical disk library, selects a platter, and loads it onto a reading mechanism.

NARRATOR: Computers analyzed the vast quantities of transmitted data and organized the information into dossiers about individuals. Unusual activity was noted in special police files.

> *VIDEO*: A sequence showing a person from eastern Europe as he goes about his daily life. Almost every aspect of it is being monitored. He talks to a woman in the street. "I can leave my paging unit at home if you open the door and if I don't have to pay for anything."

NARRATOR: During the 1970s and 1980s the West became used to accepting some controls on freedom for the general safety.

> *VIDEO*: An old lady being body-searched at an airport because she set off the metal detectors.

NARRATOR: A humiliating experience, but one that society accepted because of the need to prevent hijacking.

Cyberveillance developed fast during the troubles with Moslem terrorism.

> *VIDEO*: Police in a patrol car checking a computer screen and talking on the radio. A rotating antenna on a patrol van.
>
> A child sailing a toy motorboat on a pond holds an inexpensive radio controller with which he steers the boat. Cut to newsreel shots of Lord Mountbatten's boat, similar to the toy boat. It was blown up, killing Mountbatten, using the same radio controller.

NARRATOR: Terrorism has a dynamic of its own that destroys freedom. When terrorist attacks spread, the police find techniques for cracking down. The terrorists then attack the police or the authorities. The police then grow tougher and harsher. The terrorists claim police brutality and make worse attacks. The police spread their surveillance techniques, search innocent citizens, stop and search cars. Privacy is lost.

> *NEWSREEL*: Italy's Moro kidnapping; 1980 news headlines of the ransom demanded. A group of youths being roughly body-searched against a wall. Detectives checking computer screens. The police crackdown becoming violent.

NARRATOR: It was a delicate balance: on the one hand, effective action against the terrorists; on the other, the preservation of individual liberty.

Computers became more and more essential as police tools. Large data bases of information grew up. There was information about law-abiding citizens and much more about suspects. A constant conflict existed between the need for police information and the need to protect the citizens' liberties. When crime became worse, the balance moved toward more surveillance, more demands for information.

By the 1990s hundreds of billions of dollars per day were moved electronically over bank networks. The public used automated teller machines in the streets, and banks exchanged funds worldwide.

> *VIDEO*: Montage of bank computer usage and funds transfer. Then press headlines about wire-transfer theft.

NARRATOR: An epidemic of computer network crime developed, with vast sums being syphoned electronically from the networks into Swiss accounts. Normal police methods had great difficulty coping with this type of crime, so secret government operations were set up in cooperation with the banks to monitor electronic banking. Both the criminals and the detectives steadily raised the sophistication of their techniques.

Slowly a body of law evolved that was intended to define and protect citizens' rights to privacy when such police methods were necessary.

The more the surveillance electronics improved, the more intricate the balance became.

VIDEO: An Asian man at a Western cocktail party. The camera suddenly does a fast zoom to his ornamental tie pin. Cut to one-inch tape moving slowly. Communication satellite. A bank of 50 large reel-to-reel one-inch tape recording units. A fast zoom to a flower in a vase. A close-up showing an optical fiber on the flower stem. Flickering lights of a computer.

NARRATOR: Artificial intelligence techniques became highly developed in police work for sifting through the vast amounts of electronic data, automatically monitoring and drawing inferences.

The need for surveillance became drastically more urgent as nuclear and biochemical weapons dropped in cost and spread among third-world countries. Throughout the world computer and intelligence networks were deployed with surveillance devices, vast data banks, and AI techniques for analyzing vast quantities of transmitted data.

VIDEO: Future military satellites with cameras with 1-foot- and 20-foot-long lenses (like an astronomical telescope).

NARRATOR: Military satellites will put substantial computing power in orbit. The small camera is used to detect differences in the land photographed from one orbit to the next. The big lens photographs the differences or photographs according to orders from earthbound computers. More than 100 such satellites are controlled continuously from the U.S. National Security Agency, where staff have access to several hundred high-resolution screens. These can be switched directly to the satellite cameras or to computers controlling the world's largest data banks, where images from the satellites are stored and indexed.

VIDEO: Computer-processed Earth Resources pictures on screens. Zoom on a screen of a missile installation. The image closes on the equipment, and then zooms further to the grainy manufacturer's label on the equipment.

NARRATOR: The NASA Earth Resources satellites produce beautiful computer-processed pictures of crops, mineral resources, pollution, and other features. Vivid colors are used for coding and highlighting requested types of information. The NASA pictures have a minimum resolution of 10 feet. The military photographs can have a resolution of 2 millimeters.

The systems needed for military surveillance could benefit the civilian world as much as they benefited the military. In some countries they were designed to serve multiple purposes. Hand in hand with the deployment of such facilities it was desirable to improve privacy legislation and ensure that an appropriate constitution would protect future generations.

VIDEO: Aerial shot as if from a satellite of a car speeding around a cloverleaf. A computer flashing a cross is superimposed on the car; a digital readout is at the corner of the frame. The V and H references (map coordinates) are changing. The car parks, and a man gets out. The computer cross follows the man until he goes inside a building. A floor plan is shown with the flashing cross going down a corridor. After a brief discontinuity the cross flashes in a room and a man's voice greeting a woman, crudely digitized, is heard on the sound track.

Two officials watching such a computer screen dissolve to two people staring at a goldfish in a bowl.

NARRATOR: The need for laws to protect individual privacy had been recognized as early as the 1960s. In 1978 West Germany passed a set of laws to ensure the privacy of the individual in the age of data banks. Other countries developed and refined what the West Germans had started. The law insisted that all personal data be made secure from casual access or theft. It became technically possible to lock up data in computer systems with great security.

Throughout history, cryptography has been used to make data unreadable. Elaborate codes and code books were used. Also throughout history, brilliant cryptanalysts have worked to crack codes, often with spectacular success.

VIDEO: Shot of Pearl Harbor bombing. Shot of bombing of Coventry.

NARRATOR: Japanese coded messages saying that Pearl Harbor would be destroyed were deciphered by the United States two days before the attack. German codes saying that Coventry would be destroyed were similarly cracked, giving Churchill some tough decisions. World War II is full of such stories, and it used to be said that "there is no code that cannot be broken."

When the encoding became done by computers, however, it was

possible to scramble data so formidably that there was virtually no chance that the code could be broken. The battle between the encoder and the code breaker was increasingly won by the encoder. Computers could make data safe if careful precautions were taken.

The privacy laws insisted that data be used only for the purpose for which they were collected. Citizens were given rights to inspect data stored about themselves and to have them corrected if there were inaccuracies.

There were severe penalties for using incorrect data about people, as often happened in the early days of computers.

Modern technology enabled thorough electronic safeguards to be built.

VIDEO: Montage of a hand being pressed on a security pad. Advanced radar units. A television camera the size of a thimble rotating. Two people at home searching through an index to data on a TV screen; the machine speaks: "Voice identification required to access this file."

NARRATOR: Data can be locked up in a computer system more safely than money can be locked up in a bank vault. Administrative safeguards are well thought out today. But in the final analysis, society is dependent on a benevolent government.

What can an advanced technological society do to ensure a benevolent government? True democracies, in which there is representation of *all* adults rather than *some* adults, evolved in relatively recent history. They are constantly surrounded by pressures that could destroy democracy—human lust for power and wealth, ignorance, bureaucratic growth, the desire to mechanize procedures that limit freedom, violence, drugs, police oppression, a public living in a fantasy world created by the entertainment industry. Democracy needs care and attention if it is not to wither.

Western democracy was in fairly poor shape in the late twentieth century. Less than half the American voters bothered to vote in presidential elections.

NEWSREEL: Film of elections showing the sense of dissatisfaction with both the major candidates. A news commentator interviews people in the street.

Commentator: Are you going to vote?

Woman: What's the point? I can't stand either of 'em.

Man: It wouldn't make any difference. They'd both do the same thing.

Second Man: It's not the politicians that run this country anyway.

NARRATOR: Once in the French presidential elections, Coluche the clown secured a substantial vote, his platform being nothing more than a mockery of government and all its processes.

NEWSREEL: We see Coluche in action, clowning for votes.

NARRATOR: The lifeblood of democracy is communication. A totalitarian regime abhors the free flow of information. It wants the information channels to be controlled by the government. It tends to create channels *from* the government *to* the people. A free society needs channels from the people to everybody.

VIDEO: Scenes from the film *1984* showing the Ministry of Truth.

NARRATOR: At one time it was conjectured that electronic media might replace printed media, electronic newspapers might replace paper newspapers, videodisks and terminals might replace books. Today our view is that society needs the maximum diversity of information channels. Different media tend to convey different messages and be used for different purposes. An electronic newspaper is quite different from a paper newspaper, and society needs both.

In the Soviet Union copying machines have been banned ever since they were invented, except in high government offices. Free interchange of disks on personal computers has been banned. Networks of home computers like those here do not exist.

A free society needs the maximum diversity of lateral information channels, from the people to the people, not centrally controlled. Electronics has created important new types of information media. There need to be many different types, under different ownership and control.

To ensure the maximum likelihood of freedom in the future, we *must* ensure the maximum diversity of information media.

NARRATOR: No medium has had a more powerful effect on the democratic process than ITV, interactive television. Interactive television had its genesis in Europe in the 1970s when the first videotext systems were introduced. Ironically, Europe was extremely resistant to the use of ITV later for changing the political process.

The public took more interest in videotext when it was used in conjunction with TV broadcasting. In some countries it was used for off-track betting while the race events were shown on the same set. In some countries it was used for impulse buying in conjunction with TV advertising. Its effect on the political process began in a small way when talk shows began to solicit the views of their audience.

VIDEO: A beauty pageant taking place with the commentator building up excitement asking the home audience to vote for the girl of their choice. Fingers pressing a button on the hand-held keypad, which is connected by radio to the electronics in the TV set. The keypad has "yes" and "no" keys as well as digits. The set relays the response to a computer at the TV studio, which generates a color chart summarizing how the public has voted.

Cut to a politician holding forth on television. The moderator of the program says, "Well, that's a very interesting statement, but I'm not sure that the public would agree with you. Let's ask them." A montage of many fingers pressing the "yes" or "no" keys. Two counters running on the television screen, one saying "yes" and one "no." The moderator and politician look at them as the "no" counts climb high.

NARRATOR: At last the public had a means to answer back, to attack the politicians.

VIDEO: Moderator of a TV show: "On a scale of 5, how would you rate this?" We see a scale ranging from "1: Terrible" to "5: Excellent." Montage of fingers pressing keys. Animated faces of a viewing family anxious about the result. A rapidly growing histogram on the screen showing the result.

NARRATOR: The talk-show hosts learned how to make this type of interaction as entertaining as possible. The ratings soared. Similar techniques were used by advertisers for taking orders and a new money-making service was born. Like TV itself, it was a precocious child from the start, destined to change the grass roots of democracy.

National TV debates were arranged on any topic that touched the public. The public was asked to respond in all kinds of imaginative ways, and the responses were shown on the screen with computer graphics.

Much of the use of interactive television was informal, with a high entertainment content. On some subjects it became serious and formal. Techniques were devised for validating the responses, and these techniques improved steadily. Eventually the medium was used for political referendums.

On some issues education of the public was thought to be vital, and a final vote could not be registered unless the voter had watched and interacted for the whole program. The computer registered "attendance" and validated each vote. This tended to stop uninformed voting.

The public at last had the feeling that it had a say in affairs. Increasingly, people wanted to join in the programs themselves, appearing on screen.

Inexpensive studios were set up in many local communities. Whoever wanted to appear was shown on the local cable TV system. Viewers of the local system graded those programs according to whether they agreed with the viewpoint, liked the quality of presentation, and thought that the presentation should be broadcast to a larger community. Broadcasts with the highest ratings were shown on regional television with a much larger audience. Again the viewers graded the program, and those with the highest ratings appeared on national TV.

VIDEO: A scene with an orator rather like those at London's Hyde Park Corner, standing on a small stage in a TV studio surrounded by hecklers.

In a control room with eight monitors, a technician switches to cameras showing the hecklers.

NARRATOR: As this system grew, everyone had a chance to have his or her say on TV. Communities championed their local orators and helped them to achieve higher-level coverage. The system applied to would-be political philosophers equally with stand-up comics.

VIDEO: Scene in a bar with a TV debate going and a local orator holding forth.

NARRATOR: It became possible to be promoted quickly from debating at the local bar to national TV debates.

In addition to high-quality presentations being promoted from local to regional to national TV, there were also debates in which any person could broadcast from the local studio, whether qualified or not. A computer "draws" some of the names to appear. The phone-in radio talk shows of the 1970s became dial-up TV shows 30 years later.

Big corporations spend lavishly on their presentations but are limited in time, like individuals, and the public became highly sales-resistant to the corporate hard sell.

To ensure that television is used for political debate, one channel is devoted to this full time, in most countries, and for one prime-time hour each night *all* channels carry interactive political or news programming.

In some countries it became law that all TV sets sold must have electronics for public interaction.

VIDEO: Imposing pseudo-classical architecture of a government building.

NARRATOR: One other problem with Big Brother became very serious. Year by year the size of the civil service kept growing in most countries. Bureaucracy looks after its own. The costs of bureaucracy grew higher and higher, its value increasingly questionable.

VIDEO: Montage of a deluge of forms being filled in.

NARRATOR: One of the worst effects of the computer was that it enabled bureaucrats endlessly to increase the complexity of their procedures.

NEWSREEL: *European commentator*: Biscuit makers are entitled to a different subsidy in each country. The subsidy percentage for butter would be different from that for flour or sugar, and so the subsidy for a biscuit exporter, perhaps no more than a penny a package, would be an enormously complicated calculation, quite impossible to work out without a computer. You would not think that this is a Common Market.

NARRATOR: There seems to be no end to the trivial complexities that civil servants devised. They almost never took steps to simplify or eliminate earlier procedures but constantly devised new complexities and new forms to fill in.

VIDEO: Television comedian in 1980.

Comedian: God appeared to Moses and said, "I've got good news and bad news." (laughter) "The good news is that I'm going to part the waters of the Red Sea so that your people can walk across. Then when the Egyptians follow, I'll release the waters so that they'll all be drowned."

Moses said, "It sounds incredible. What's the bad news?"

God said, "You have to fill in the environmental impact forms!" (loud laughter)

NARRATOR: In industry there was constant pressure to cut costs. Competitive threats demanded greater efficiency. There was a constant drive to cut procedures to those that directly enhanced profit. Computers were used wherever possible to eliminate paperwork. The term *paperless office* became common.

VIDEO: Montage of terminal screens. Images on the screen like those of the Macintosh office.

NARRATOR: While competitive corporations struggled to cut administrative costs with office automation technology, government departments let the bureaucrats protect and multiply their jobs, increase the form filling outrageously, and use the computers to complicate instead of simplify.

One part of the economy struggled for higher productivity while another part squandered the gains.

NEWSREEL: Employee in a government setting with a flag in the background.

> *First Employee*: The people from UGNS come down to our audit department and they just give 'em the runaround. They are told to wait for an hour. They think the boss's busy. He could see 'em if he wanted. He's demonstrating his power. The ding-a-lings think they're getting somewhere, but they really aren't. They go from floor to floor, from person to person. "Fill in this form, fill in that."
>
> *Second Employee*: Every now and then there's a protest about the system. So they set up a task force. They love task forces. There might be 20 people on the task force. Nothing ever gets improved. It's all political.
>
> *First Employee*: It's a job. I get my paycheck. What can you do? Play the stupid game.

NARRATOR: The development of artificial intelligence techniques gave rise in the late 1980s to a powerful new form of data processing. Complex procedures were expressed in terms of rules. A computer could shift through large numbers of rules, draw inferences, and produce results. To change the behavior of the system one changed the rules. This technique was used for building expert systems that were to become vitally important in most human activities—oil exploration, financial investing, control of automated factories, medical diagnosis, car maintenance, even helping housewives to look after house plants.

Rule-based systems changed administrative procedures. Procedures were expressed by computers. The computers were told when human judgment was needed, and to the greatest extent possible procedures were automated. The Digital Equipment Corporation was one of the first to use rule-based systems for configuring complex orders, helping sales reps, controlling operations in the factories, delivery planning, and so on.

It became clear that most procedures and laws could be expressed in terms of computable rules. Legal aid systems were built to give people advice on taxes, setting up companies, divorce, and so on.

As expert systems were applied to the law, they showed that many laws were too vague or ambiguous for computer processing. The computer revealed imprecision in the legal system. Laws ought to be drafted so that they are computer-processable as far as possible. Computers were programmed to search for legal precedents. Such a use of computers could greatly lessen the costs of legal action and relieve the bottleneck in the courts. Nobody suggested that the computer should replace the lawyer or the judge, merely that it should speed up the logical part of their job. Computer-aided drafting of contracts led to automated advice on the avoidence of litigation.

Many lawyers, however, rebelled against too much precision in contracts.

VIDEO: Scene in a lawyer's office.

Lawyer: The term *fiduciary responsibility* does not have an exact meaning in this context.

Client: Then why did you put it in the contract?

Lawyer: If *they* are trying to pin us down, I like the terms to be as precise and narrow as possible so that we know exactly where we are. If *we* are trying to get *them* over a barrel, I prefer to use terms more open to interpretation because then we have much greater scope for threatening litigation at a later time.

NARRATOR: Efficient corporations automated their rules and procedures as fully as possible. Government, however, was often the last to streamline its operations in this way. Designing the tax laws and other laws for computer processing would have saved everybody time, expense, and emotion.

In most countries the percentage of white-collar workers working for government steadily increased. Taxation rose accordingly. Various politicians running for office promised to cut the civil service.

NEWSREEL: Reagan, Thatcher, and a politician of 2005 promising to cut the budgets of government. Details of how the budgets increased, not decreased.

NARRATOR: On isolated occasions a politician managed to cut the bureaucracy, but most had no success at all. The civil service had its own defenses. It continued to grow ever larger and more expensive.

Popular feeling grew stronger against government—any government, all government.

VIDEO: Car bumper stickers saying "Shoot a Bureaucrat a Day."

NARRATOR: It became perceived that no president or prime minister was capable of dealing with the bureaucracy. It just grew and grew, destroying wealth, using endless computer time, uncontrollable.

VIDEO: A montage of form filling, EEC regulations, civil servants in Washington and Brussels, computers, forms getting longer and longer, infuriated businessmen, a mile-long line of trucks at a frontier post in Europe, civil servants wearing carnations and drinking brandy in expensive clubs.

NARRATOR: Social agitation about taxes and bureaucracy grew stronger. Many local efforts succeeded in cutting local taxes.

NEWSREEL: Proposition 13 oratory in California in 1978. Property taxpayers' strikes in London, 1990. Angry crowds with placards. Police barricading the crowds.

NARRATOR: In some areas disillusion turned to violence. Some of the public could see no method of controlling the civil service other than violence.

NEWSREEL: Windows being smashed with placards. "Kill the Bureaucrats" stickers. Angry crowds. Fire engines. Wreckage of a burnt-out computer center in a government department.

NARRATOR: As the civil service increasingly felt itself under attack, it hardened its attitudes and increased its protective measures.

It was countries with full use of interactive television that first dealt effectively with this cancer-like problem. The first step to controlling bureaucratic excesses is to control the budget of government. The television orators built up a public fury about the situation, showing one example after another of government waste and nonsense. They demanded that a ceiling be placed on government spending. The television debates argued about what the ceiling should be.

NARRATOR: (over newsreel footage) A new prime minister in Australia was elected on the platform of limiting non-defense-related government spending to 5 percent of the gross national product. Bureaucrats were allowed three years to trim their budgets to this level. It then became a constitutional ceiling on government spending.

The television orators then waded into state and city governments demanding that they also impose a ceiling. They appealed as graphically as possible to the taxpayers with television films showing gross government waste and idiocy of many different types. The public reacted strongly where they had interactive sets.

NARRATOR: A budget ceiling is not enough to control harmful bureaucracy because with today's computers there is no limit to the complexity of regulations that can be devised. Government can impose regulations with little cost to itself by making citizens fill in "forms" on videotext screens. The information is transmitted to government computers so that civil servants do not themselves have to process the forms.

Australia was the first country to pass a Freedom from Harassment Act. This was designed to control various types of bureaucratic harassment. It makes it possible for any citizen or body to challenge the necessity of a form-filling procedure that they regard as pointless or too much work. An independent tribunal reviews the procedure and may demand that it be changed. Civil servants responsible for harmful procedures are censured. This has the effect of greatly reducing the bureaucratic complexity that government imposes on its people. Television commentators conducted an onslaught on burdensome form filling.

Driving licenses and other documents did not have to be renewed periodically; they were issued for life. People on welfare did not have to line up for payments; payments were transferred electronically to their bank, and they could display their account on their home TV screen.

VIDEO: Shots of documents showing their title: "Freedom of Information Act," then "Freedom from Harassment Act."

NARRATOR: Whatever the current laws and controls, they could quickly be changed. So in the final analysis, freedom from Big Brother depends on having benevolent government. A vicious government has available to it today an overwhelming set of machinery for surveillance and harassment of the public and an overwhelming capability to crush opposition and maintain itself in power.

How do we ensure a benevolent government? The political constitution of a country is all-important in a computerized world. The constitution must be designed to eliminate abuses of freedom of any type. The public must all know and understand the constitution. It must be taught thoroughly to schoolchildren. The media, especially television, must place a high value on the constitution and be alert to any possible infringements of it.

VIDEO: A block labeled "Political Constitution" is placed on one labeled "Basic Ethics" (Fig. 4.1).

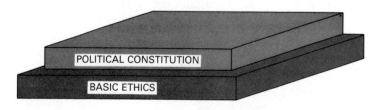

Figure 4.1

NARRATOR: The first building block of our value system is the moral and ethical code. The second building block is the political constitution.

A political constitution needs to include a bill of rights protecting citizens from encroachments on their political liberty, reductions in their autonomy, unjust invasions of their private lives, and unnecessary government harassment. When public officials exceed the authority and power of their constitutionally appointed offices, a well-designed constitution must be able to provide remedies.

Communication is the essence of democracy, and the technology of communication has changed beyond recognition in the past century. Like all powerful technology, it can be used for good or for evil. It can be used to inform, educate, and allow citizens to make their views felt. It can be used to oppress, harass, and destroy privacy. It can be used to distort reality or to drug the public with images of a fantasy world.

A modern constitution must reflect the power of modern technology.

VIDEO: Montage of small satellite dishes, electronic bugs shown earlier, fingers touching a television touch screen during a talk show and the image saying "thank you," a telescopic lens on a satellite shown earlier, a computer image of speech analysis.

NARRATOR: Democracy and the freedom it brings can be destroyed if society has to battle with an excessive degree of violence, drug abuse, crime, or terrorism. These elements bring the need for surveillance and restrictive police measures.

Democracy is endangered if the self-governing citizens do not understand and discharge the obligations of citizenship, and particularly if corruption or abuse pervades the officials of a society. A sense of basic ethics needs to pervade a society if democracy is to work well. This is why *basic ethics* is shown as the bottom block of our value system. Without this, the freedom decreed in the constitution will be impaired.

VIDEO: Cracks appear in the block labeled "Basic Ethics" until it crumbles. The block labeled "Political Constitution" falls through it and breaks.

NARRATOR: The value system for building moral and ethical behavior *must* be taught in the schools.

VIDEO: Repeat footage from Installment 3 showing school teaching of society's basic value system.

NARRATOR: For over two centuries the U.S.A. has been a predominant example of a country that takes a well-articulated constitution seriously. The Constitution is understood and acted on by the media and is taught in schools.

Various possible abuses of the U.S. political system have been nipped in the bud because the media and the power structure of the country are alert to the Constitution. As long as unconstitutional trends are stamped on when they start, a country is likely to remain free. Freedom in an electronic age requires alertness.

There must be complete freedom of interchange of information among the people. Responsible television broadcasting on political issues must be mandated. Uses of interactive television must be a political right. There must be protection of the individual's rights of privacy and freedom from harassment.

Modern media, especially television, are the cement of democracy. Television linked to computers can be the most powerful instrument of oppression or the most powerful means of democratic debate. It can be used to spread violence, it can be used as a moronic opiate, or it can be used to spread enlightenment. It is far too powerful to be left to the whims of commercial advertising. *A society that uses its television irresponsibly is a society that will pay an appalling price*.

> *VIDEO*: A final extended montage showing participatory democracy at work: A scientist arguing for better crop management. A national tax referendum. Fingers pressing the keypad of the home TV system. A schoolboy asking for another cricket pitch at his local sportsground. Computer graphics forming on the screen. An old lady campaigning for the restoration of a nineteenth-century street in Melbourne. A campaign for more money for the development of expert systems for international politics.

NARRATOR: The sequence gives an impression of political activity at every level of an intensity such as has never been seen since the days of fifth-century Athens, when it was the right of every male citizen to attend the Assembly of the Citizens—the supreme sovereign body—and propose any motion he wanted to propose. Passion has returned to democratic politics—so unlike the mechanical TV commercial posturings of an earlier era. A modern constitution must reflect the power of modern technology.

> *Titling*: A MODERN CONSTITUTION MUST REFLECT THE POWER OF MODERN TECHNOLOGY.

Installment Five

DRIVING FORCES

PRESENT-DAY AUTHOR: We are letting loose ever more powerful magic from the crucible of technology, power that we barely know how to cope with. New technology can be used for good or for evil. The benefits, if used well, can greatly improve the human condition. The dangers could be catastrophic in many ways. It is essential for a philosophy of technology to emphasize how positive effects can be maximized while creating controls to prevent negative effects. We are likely to evolve to better forms of civilization, but as with all evolution, there will be negative effects to be dealt with, and some dead-end branches.

The curves we draw of technological change grow geometrically, not linearly. The technology doubles in power or capacity in a given time, then doubles again in the same time, doubles, doubles, and goes on doubling. It is like a cart carying a bunch of children going down a hill. First they push it very slowly—one mile per hour, then 2 miles per hour, then 4; it picks up speed: 8, 16, 32 . . .

We are involved in a wild adventure—the wildest adventure in the history of humankind so far. Where will it end? Is anyone in control? Can we steer it? Or are we in a runaway vehicle with no real steering mechanism? A vehicle that travels faster every year?

The question of how we steer technology, or whether we can steer it, is an interesting one. Scientists continue to do research. There is almost no occasion when a moratorium on research in a given area has lasted for long. In spite of alarm over nuclear weapons, research in nuclear physics continues with ever greater expenditures on particle-smashing machines. Concern about unemployment has no effect on research to create better robots and artificial intelligence.

Sometimes it seems that technology is moving so fast that our institutions cannot keep up. But soon it will be moving much faster. The era of artificial intelligence has barely begun. The era of robots or genetic engineering has barely begun. Computers are being used to design much better computers. The explosive growth in computer power speeds up all other technologies. How can the institutions of today's society—government, jobs, schools, the family—adapt quickly enough?

If the children with the cart on the hill have difficulty steering it, that does not matter much at 4 miles per hour. At 8 miles per hour it becomes more entertaining. At 16 it is exciting. At 32 miles per hour it is dangerous; at 64, possibly lethal.

Many technologies are still at the entertaining or exciting stage. Some cause temporary solvable problems. The effects are sometimes irreversible or very difficult to reverse. Some effects are long-term, and we have difficulty understanding the outcome. No one understands the long-term outcome of computers with artificial intelligence, for example.

Technology is changing our lives much more drastically than any other force. Politics, especially bad politics, as in oppressive regimes, can change people's lives seriously, but in the long course of history, political excesses become changed, whereas technology remains ever present. In view of the power of this force, we badly need a subject in our universities called the philosophy of technology. This subject barely exists today. It is quite different from philosophy of science because it is concerned with people, sociology, ecology, control of war, the political process, the law, adapting the Constitution, and other such matters.

In doing the research for this series I scoured the bookshops of major universities for relevant material. There seem to be almost no books on the philosophy of technology that portray future technology realistically. There were many negative diatribes against technology by liberal arts professors. There were books about the future from an engineer's viewpoint. Experts in different fields who are able to envision the future potential of their field regard it with awe but cannot translate it into its full social consequences. In our universities the philosophy of technology is a nonsubject.

I believe that we can make some choices and that historians of the future will look back on the present time as being a very pivotal decade in changing the course of society's evolution.

The articulation of a philosophy of technology and civilization is crucial for the evolution of humankind trying to become the master of our own fate.

VIDEO: The same scene-changing sequence as before is used to slide from the present to A.D. 2019: a rushing sensation like flying very low at extreme speed through a mountainous valley, but the earth and hills look like endless etched microelectronic circuitry. Electronic music is used, evocative of time travel.

VIDEO: A room in the narrator's home filled with highly polished oak furniture made in England in the seventeenth century.

NARRATOR: The first flickerings of the industrial revolution began in England only twelve generations ago. That seemingly innocent beginning lit a fuse that was to cause technology to race with ever increasing speed along the paths we know today. Our seventeenth-century ancestors would have reacted with stark horror to a vision of the world in which we now live comfortably. They might have called it a pact with the devil. We would probably be equally startled if we could see society twelve generations into the future.

The furniture in this room was made before it all began—before we had invented mechanical looms or steam power or paper money.

It is endlessly fascinating to speculate why it happened then, why at that particular place and that particular time. Men had built machines long before then.

> *VIDEO*: Pictures of Archimedes' screw, the water clock of Ctesibius of Alexandria, Hero's odometer, Leonardo da Vinci's screw-cutting machine (or other ancient machines), the ancient clock mechanism in Salisbury Cathedral.

NARRATOR: Even power-driven devices were built long before the industrial revolution.

> *VIDEO*: Pictures of the Roman mill plant as Barbegal; old Dutch windmills.

NARRATOR: By the early thirteenth century, windmills had spread through most of Europe and in Holland were used for pumping water. Hero of Alexandria built crude steam engines 200 years before Christ, but nobody ever imagined that they should be employed for useful purposes.

> *VIDEO*: Pictures of Hero of Alexandria's aelopile.

NARRATOR: Perhaps the most important difference in eighteenth-century England was the realization that money could be made from technology. The entrepreneur may have been more important than the scientist.

The industrial revolution changed Britain. But it was not only a technological change. Alongside, there was a change in people's material expectations. Gradually, more and more of the population wanted more money, more goods, more services.

James Watt, Arkwright, Hargreaves—the men who made the early industrial revolution—have usually been thought of simply as inventors. But they were also entrepreneurs who had an understanding of how the public could be persuaded to want more goods. Henry Ford I was not only someone who revolutionized automobile production. He also had a vision of a population of car owners. Before him, to own a car was mainly the prerogative of the rich; after him, it was possible to conceive of a nation with a car to every household. Bob Noyce and Gordon Moore similarly had a vision of ever better chips taking computer power into every nook and cranny of society.

The persons who lit the fuse of the industrial revolution were more like Henry Ford and Bob Noyce than like the traditional eighteenth-century scientists. Once lit, technology improved technology. Growing mechanical skills fed on themselves. The machines became better because of a better understanding of scientific principles.

VIDEO: A montage of steadily improving steam engines and other machines in motion.

NARRATOR: War, or the fear of war, has often been one of the strongest driving forces to create more advanced technology.

VIDEO with narration: An explanation, in a museum, of how there were no metal cylinders in steam engines until the very end of the eighteenth century, when precision casting capable of making the cylinders was developed. But this precision casting was developed only because of the military need to make accurate cannons.

We see the inadequacy of aviation before World War I.

NARRATOR: If there had been no war, it might have been years before civil aviation developed as it did.

It was because of the development of radar in World War II that microwave telecommunications could be developed in the 1950s and 1960s. It was because of the atomic bomb that there were nuclear power stations.

VIDEO: Montage sequence of military technology. The first tanks in World War I, the first radar, the atomic bomb, ICBM launch, a space weapons control room in 2019, a supersonic cruise missile flying very low through the valleys of a mountain area, a nuclear-powered X-ray laser unit in orbit.

NARRATOR: The advanced technology of the military soon finds its way into consumer products.

VIDEO: Shot of military fighter dissolves into a scene of a child playing with a videodisk with a similar fighter on the screen.

NARRATOR: This child's toy uses the same microchip technology as the fighter itself.

Fear and greed are two of the most effective human motives. In the 1960s fear of Russian space technology led to the Americans' putting men on the moon.

NEWSREEL: Newspaper posters in London, 1957, saying "Red Moon over England." Shocked reaction to *Sputnik* on U.S. television.

NARRATOR: The reaction to *Sputnik* was minor compared to the reaction to the terrible events of April 10, 1999.

VIDEO: We see a series of still photographs of Middle Eastern cities from the air. Middle Eastern city sounds accompany them.

Shot of a photographic surveillance satellite of 1995 with a large-diameter 10-foot-long telescope lens.

A rapid sequence of people, human faces, men in military uniform, children playing, tanks, a crowded bazaar, communications antennas, all photographed from above.

A rapidly cut sequence of nuclear explosions.

Billowing clouds photographed from above. An unnatural abrupt cut in the sound track to total silence.

Aerial shots of complete devastation, as in Hiroshima. These last for 20 seconds, with no sound track.

NARRATOR: Pictures we have grown up with. A war to end all wars. Which is how the other two great wars of the twentieth century were described; but this time it had better be true.

NEWSREEL: News film of the Middle Eastern war of 1999. The early provocation, the border incidents, the attack on the oil fields, the first atomic attacks.

NARRATOR: Perhaps it was inevitable. We needed to have one nuclear war. To see it on the high-fidelity wall screens.

Fortunately, the madness stopped before it spread beyond one subcontinent.

NEWSREEL: We see the reactions to the Middle Eastern war. Worldwide panic. Stock markets crash. A mad rush to buy gold. Housewives strip supermarkets bare in a matter of minutes. Fights at gas stations as stocks run dry.

NARRATOR: 1999 was Australian election year. (The narrator describes the situation over news film of the time.)

Australia was already panicking because of the war in southeast Asia. Vietnam was already in a state of semi-war with Indonesia.

On top of that, as the election campaign was about to start, a cache of unauthorized plutonium was discovered—presumably collected for some terrorist purpose.

The liberals had just elected a new leader, Karl Weinberg. He immediately changed his manifesto and went to the polls saying that above anything else Australia must be defended.

NEWSREEL: We see some of the Weinberg election campaign. His personal charisma. A sunburnt, athletic, Australian Disraeli.

Weinberg: Australia is in deadly peril.

This is a vast continent with enormous mineral wealth—a sitting plum for aggressors.

Because of the chaos in Southern Africa, Australia will be one of the world's most important suppliers of diamonds and strategic minerals such as uranium, and cobalt and titanium, which are vital for the defense industry.

To defend ourselves, to exploit our minerals, and to further open up our northern territories, we must make Australia a crucible of high technology.

NARRATOR: A crucible it became.

Weinberg commented on Anglo-American democracy's unpreparedness for war—for World War I, for World War II, and now for World War III.

In contrast, Switzerland, as early as the 1970s, was prepared for an all-out nuclear attack and would have stood a good chance of surviving it with the vast majority of its population intact. Switzerland now, in 1999, is sure to be safe. Australia, Weinberg said, must become a Switzerland. "We must make our country safe."

He described how new towns should be built, designed to take full advantage of current technology, which would survive anything other than a direct nuclear strike. He showed how much of the population could be rehoused over a period of ten years in nuclear-protected dwellings. He promised to build new cities on the north and west coasts, near magnificent beaches, but these would be different cities— beautiful, functional, protected, and wired together with optical fibers that could relay prodigious quantities of information.

NEWSREEL: Weinberg election speech.

Weinberg: The first duty of a government has always been to protect its citizens. Australia in the nuclear age is dangerously unprotected. We need new cities, hardened factories, secure communications, and a banking infrastructure that, like Switzerland, would survive catastrophe.

NARRATOR: He proposed Project Hermes: the creation of a computer and telecommunications network, involving every dwelling, every factory, every installation of all sorts, designed both to give Australia defenses with immediate alert to any hostile activity and at the same time to create the world's most advanced information society, affecting all aspects of work and leisure—all to be completed in ten years. In the horrified, traumatic reaction to the events of 1999, Weinberg was given all the support he needed. He was returned to power with the largest majority in Australian history.

VIDEO with narrative: We see the development of Project Hermes. We see the surveillance system, so much more all-embracing than earlier American ones. Every plane, every boat, every military vehicle, every gun, every piece of plutonium carries a transmitting device to fix its whereabouts. Any unidentified plane or boat is immediately registered, its position plotted, and is then challenged; if the challenge is not answered, action is taken. The course of any approaching missile will be plotted and will provoke instant retaliation. Any unusual chemical detected in the atmosphere will trigger countermeasures.

NARRATOR: The systems of protection used giant satellites, sensors, networks of computers, and artificial intelligence. Small packets of data could be transmitted from tiny portable devices anywhere and relayed over the networks. There was complete surveillance of coastal waters. Any boat or plane must transmit coded information continuously or invite immediate retaliation. The air is constantly analyzed for unexpected chemicals and organisms.

Artificial intelligence techniques for analyzing surveillance and intelligence information required giant sixth-generation computers, some of which employed 10 million inference processors operating in parallel. Supercomputers in different locations exchanged prodigious amounts of information over the optical-fiber networks.

Unmanned cruise-like planes were continually in flight to combat any invasion.

VIDEO: A cruise-like plane flies at supersonic speed very low through a mountain valley. The systems in the following discussion are then illustrated.

NARRATOR: In orbit, patrolling American attack satellites could flash an intense X-ray laser beam for a microsecond. Surveillance satellites scanned constantly for possible targets for this lethal laser flash.

Much larger laser beams could be directed from earth to orbiting mirrors positioned with minute precision by computers. The earth laser stations consumed more electricity than a large city for the instant during which their beam was switched on.

Intense biological research ensued on protection from nerve gas and biochemical attack. Systems were built to react instantly to any such attack.

VIDEO: A giant aerosol spray with a dozen nozzles squirting vapor in the sunlight.

NARRATOR: Most military research has given substantial civilian fallout. In this case Weinberg was determined that the beneficial byproducts should be planned from the start and should determine much of the design of the facilities.

Project Hermes may have been triggered by military needs, but it provided the "wired society" benefits about which so much had been written.

Many of the new Australian cities were physically isolated and so were more dependent than older cities on electronic connections, which provided them with education, culture, links to medical specialists, entertainment, and general interaction with the older communities.

The electronics of Project Hermes changed the very nature of democracy. It provided all manner of services.

As a result of it, every citizen is in two-way communication with central control and with a hierarchy of local control centers. This communications network can be used with ease for social purposes. What has happened to 85-year-old Mrs. Smith? She has not checked in for several days. Quickly we discover—either from her "neighbors," who could be many miles away, or her local control— that she has had a fall. Quickly help is on its way. The disabled can ask for help; the energetic and charitable can offer help.

Project Hermes was from the start a holistic design, integrating defense, social needs, and economic development.

The architects asked the question, What are the most attractive features that one would like in a city built to take fullest advantage of advanced technology? The need to put factories and houses at least partially underground gave the ability to create a park-like appearance for much of the city. Multimegabit communications, wall screens, and the latest computers made it possible for most people to work within walking distance of home. The shopping malls and precincts were built to be free of traffic except for early-morning deliveries.

VIDEO: We tour the new cities, observing various ways in which houses have been built, usually cut out of real hillsides or built into artifically created hillsides, with roof garden; solar generator; an aboveground part, made mainly of glass and containing nothing essential to the running of the house; then the belowground office and learning rooms with their screens and terminals, the automated kitchen, and so on.

The underground houses are silent and full of light. Large windows and prisms direct the light to all parts of the house. Sometimes the prisms and diffraction gratings are designed to give a blaze of rainbow colors. The prisms rotate, following the sun. As the camera pans, the viewer is struck by the richness and beauty of the furnishings and drapes, sculpture and artifacts. It is an elegant environment filled with the ornate designs of an age of complexity. Robot-woven textiles and intricate paneling and inlays glitter with the spots of light from computer-designed chandeliers.

The house plants and flowers would have amazed indoor gardeners of an earlier era, because now they were fed and watered under computer control, optimized to produce the best blooms. Hydroponic garden systems were integrated into the architecture.

No home, however close to its neighbors, is bothered by its neighbors' noise. Furthermore, these cities, unlike the cities of the past, do not destroy productive land. Their top level is gardens, fields with sheep, and parks. In the scenes with sheep and flowering bushes we can hear distant sounds of a Mozart flute concerto, and someone practicing on the piano.

NARRATOR: The earlier passion for jogging had evolved into a much more widespread desire for walking, and the design of the cities assumed that most people enjoyed a half-mile walk to work, to pubs, or to shops. Heavy shopping goods were delivered to homes. The areas designed for walking were free of traffic.

The houses used microcomputers for many functions. Rainwater was collected on the roof and stored. Water was recycled, purified, and reused for purposes other than drinking. Sunlight provided the energy for air conditioning. The underground units were dry and toasty.

NARRATOR: (over video) It was easier to build new, dispersed, semi-underground cities in the uninhabited north and west of Australia, which is where most of the minerals are. In order to entice the Australians there from Sydney and Melbourne, the cities had to be made very attractive. Large incentives were given to architects, designers, civil engineers, and transport engineers to create a new living system from nothing.

When it became apparent that Weinberg's plan was working, Australia became an extremely attractive country in which to be. At a time when nuclear war seemed possible, many people wanted to move their families to an environment designed to survive. To some it was the only meaningful insurance policy.

Weinberg was determined to use this bait of safe living to attract "the best and the brightest." Top scientists, artists, and other people of achievement were offered the chance of instant Australian citizenship. Huge financial inducements were given to anyone who undertook desert reclamation.

As English-speaking entrepreneurs around the world saw what was happening, they flocked to Australia. Weinberg's taxation policy did everything to encourage them. This was in stark contrast to earlier decades, when Australia had unwittingly tended to drive out its entrepreneurs.

Entrepreneurs generate wealth and jobs. Wealth tends to generate more wealth.

Many multinational corporations decided to invest in the new Australia, which needed their latest technology. They built vast networks of factories with the latest robot control, unfettered at last by unions trying to prevent them.

VIDEO: Helicopter shots of ultramodern semi-underground factories with vast solar panels, clusters of two-bladed wind generators, and complex antennas. Many houses are within walking distance of the factories. There are gardens everywhere with flowers growing in hydroponic beds.

NARRATOR: The glorious coastline for 200 miles to the north of Perth developed from nothing into the California of the future. "Crucible of high technology" became the slogan for a decade.

Much of the technology with which Australia built its "wired society" was foreseeable a decade or more earlier. As so often before, it was war or the threat of war that boosted massive research and development of technology. The urgency of this was made clear by the images of nuclear horror played so vividly on the wall screens. How could we stop war from happening again? Any price was worth it.

Just as the fear of Russian space power led to Project Apollo and put men on the moon, so the fear of nuclear attack made Australia vote for its wired society, which led to Australia's current prosperity and cultural achievements.

The society that has resulted from this crucible of high technology would probably have come to pass without the war of 1999, but it might have taken much longer. But in a sense it was a pact with the devil.

The technology that gives us civilizations also gives us weapons. To make more advanced weapons we strive for advanced technology, which in turn has given us the machines without which life would be unbearable drudgery, our culture decimated, and the world population of 8 billion unfeedable.

War and defense have caused major innovations in technology. At certain times, usually times of crisis, government (in the West) initiates great projects like the Manhattan Project, the moon shot, and Project Hermes.

But most new technology does not come from government. Most

inventions and patents come from relatively young corporations and individual inventors. Much of the most important and most original new technology since the industrial revolution has come from such sources. The James Watts of today are vital to the technology of tomorrow. Their driving forces have not changed. The stage is different, but the types of actors are the same.

> *VIDEO*: A sequence of portraits of industrial innovators: James Watt, Matthew Boulton, Henry Ford I, Bob Noyce, among many others.

> *NEWSREEL*: A great industralist of the 1990s who had a similar vision to that of Henry Ford I: one in which every household should be equipped with various household robots.

> We see a production line 5 miles long of robots manufacturing other robots.

NARRATOR: A small number of men are destined to play a massive role. The days of the railroad robber barons are over. But new billionaires are emerging. We do not often recognize future billionaires at age 40, but they are among us—far more of them than in earlier eras.

The urge to experiment is innate in scientists, but most of the exploitation of technology has been driven by the demands of the marketplace.

> *VIDEO*: We see a woman in an automated kitchen of 2019. She unloads her groceries into the compartments of the robot cooker. The packets are filed away by the machine, some in deep-freeze storage, some in refrigerated storage, and some at room temperature. She switches on the machine's video screen and examines the menus it offers to cook with its current stock. She selects a meal by touching the menu. The screen shows her a color picture of the result. She agrees and enters an eating time. The machine then selects certain packets, moves them to its workstations, and starts preparing the meal.

NARRATOR: The forces of the marketplace have resulted in great innovation and a great rise in standards of living.

Essential to marketplace activities is advertising. Advertisements constitute a major element of the driving force.

> *VIDEO*: It is easy to make out-of-date advertising look funny. A 2019 audience is shown reacting with glee to a showing of old advertising: starting with nineteenth-century posters, then early advertising film (the first advertising film was before sound), and then the more sophisticated film of the 1930s, for example, Henry Ford's advertising

films with catchy music. The audience is shown laughing at the television advertisements of the 1980s.

NARRATOR: In some cases the advertising campaign precedes the creation of a product. If the advertisements look good and market research shows that the public will demand what is advertised, the product is made to match the advertisement.

VIDEO: We see scenes in a product-planning meeting of a large corporation. The computer screen displays the latest survey results: "Ease of keeping it clean rates 82. Much higher than fuel economy, 46. Choice of style rates 73, but that varies strongly from one country to another." "It looks as though we need a modular design that enables us to change the style for different countries."

NARRATOR: The cleverer the marketing and advertising organizations became, the more they were able to persuade people to buy goods they neither needed nor really wanted.

VIDEO: A quick selection of clever advertisements for products such as time-release deodorants, unnecessary vitamin and drug products, a voice-activated catport (so that only your cat can get in, and not the local tomcats), a topiary robot clipping a large bush into the shape of a kangaroo.

We see a garbage truck labeled "Trash Unlimited."

The excesses of the TV commercials during the final years of network TV: greed, one-upmanship, unnecessary gadgetry, planned obsolescence, and waste.

NARRATOR: However much material wealth increases, it never seems to come up to material expectations. But there were people to say, "Enough is enough." Ever since the industrial revolution, there has been an antitechnology movement.

VIDEO: The nineteenth-century Luddites.

NEWSREEL: Protests against Henry Ford's first production line and 1980's unionists protesting against automation.

Back-to-nature movements. Protests in Britain about highway building. Demonstrators lying down in front of bulldozers. Future protestors smashing the landscaping robots in city parks.

We see an IBM advertisement in Britain, 1980, saying "You can't stop time by smashing the clocks."

The sheepshearers of Australia protest in 1996 against robot sheep-shearing.

NARRATOR: As technology developed, so did the antitechnology movements. Certain third-world countries deliberately turned their back on Western technology.

VIDEO with narrative: We see something of Burma in the 1980s and hear reactions. We see Iran in the 1980s.

NARRATOR: In one first-world country, there was a conscious political decision to "stop the clock": New Zealand in the 1990s had plenty of food and enough technology to satisfy basic needs, with plenty of space for its small population. In spite of the great beauty of the country, its energy self-sufficiency, and its abundance of food, the attempt failed. Many people with intelligence and initiative emigrated, especially the entrepreneurs. Unemployment soared. The economy went into a tailspin.

Technology periodically makes errors, sometimes bad errors, as the protesters eagerly point out.

NEWSREEL: A quick sequence of technological disasters: The thalidomide children: a series of stills showing them first as babies, then at various stages of growth.

The Chernobyl power station, a Geiger counter ticking furiously on the sound track.

Newsreel footage of the Vaiont Dam collapsing in Italy (October 10, 1963).

Shoppers in Tokyo with gauze pads over their mouths. Stands in the street, where pedestrians can put a coin in the slot and breathe in oxygen from an oxygen mask.

NARRATOR: There have been catastrophes in the development of technology. In some cases entire lines of technological development have been abandoned.

NEWSREEL: The *Hindenburg* dirigible exploding in flames, people falling to the ground.

NARRATOR: Technological progress is somewhat like evolution. There are errors and blind alleys. Sometimes entire spurs of technology are scrapped. There is trial and error. In the long run it does lead somewhere.

As any computer programmer knows, complex systems have bugs in them. The bugs can eventually be found and corrected.

NEWSREEL: Newsreel footage of Galloping Gertie, the U.S. suspension bridge that snaked for hours in a violent wind and then collapsed.

Newsreel footage of a Cape Canaveral launch in January 1980, when the rocket turned around and accelerated vertically downward into the lauching area with a vast explosion.

NARRATOR: In the rapid growth of a technological society it is perhaps inevitable that mistakes are made, sometimes on a grand scale. People eventually take corrective action.

However, as technology becomes more powerful, the mistakes become bigger, and as it becomes more complex, especially with artificial intelligence, it becomes more difficult to debug.

NEWSREEL: We see the complex control room of a giant breeder reactor in 1998. The controllers are in panic.

The siren is blowing. Excited Frenchmen are screaming. We see a technician checking charts at the screen. Curves on the charts are dipping into a zone colored red. Alarm lights are flashing. Men run in panic. Huge machinery is shaking dangerously. Police with loudspeaker vans are evacuating a town. In a river the water is boiling. Dead fish. Signs in French by the river, warning all citizens to keep away from the water. Worldwide alarm about radioactive fallout of much longer half-life than that from Chernobyl, 12 years earlier. Woman in supermarket testing vegetables for radioactivity with a Geiger counter.

In Brazil a furious debate followed. Brazil was operating a French breeder reactor. Should it be closed down? No. Brazil needs the electric power. Antinuclear demonstrators are roughly arrested by the police.

NARRATOR: Once mistakes or dangerous alleys are recognized, they can be corrected or avoided. Often the mistakes themselves lead to better approaches or new technology. The *Hindenburg* disaster gave impetus to the development of airplanes. The thalidomide children caused drug safety controls and increased work on prosthetic limbs.

Chernobyl and then the breeder reactor disaster in France was a huge encouragement to the development of fusion power, so that fission power, with its dangerous radioactivity, was eventually made obsolete. The fusion reactors, much smaller than originally envisaged, made energy abundant once again. Along with solar energy and biological fuel sources, we have clean, nonpolluting power.

Sometimes the mistakes are less obvious at the time, persist for decades, and are then very expensive to correct. The destructive impact of traffic on towns, for example, was tolerated for far too long and could only be changed over decades.

VIDEO: Heavy trucks thundering through a small English village. Fumes and traffic jams. The noise and chaos of traffic in Naples.

Twelfth-century sculpture eroded beyond recognition from fumes. Cracks in ancient frescoes.

NARRATOR: It is barbaric that we allowed so much of the great art of centuries to be destroyed by traffic.

VIDEO: Diners in an open café frowning at the noise. An ambulance. A pedestrian dead.

A chart of traffic fatalities.

NARRATOR: A basic principle of town design should have been that traffic and pedestrians are separated. Communities should be beautiful places to walk, sit, and socialize.

VIDEO: We see the narrator in her garden city. Clean elegance, electric cars, landscape robots at work, no roar of traffic, no diesel fumes, no smoke. Seats in the flower gardens. Sculpture everywhere. A city in action, but clean and human, rich in its complexity.

NARRATOR: But can we steer science? Can we anticipate and avoid its problems?
   Until recently there has been little success in consciously directing the long-range evolution of technology. The steering mechanisms have to a large extent been the demands of the marketplace. These do little to anticipate the bad effects, although they certainly demand corrections when the bad effects are visible.
   One of the problems of artificial intelligence is that very intricate errors—bugs—can occur, and these sometimes manifest themselves when least expected.

VIDEO: The year 2000. A jumbo jet carrying 1000 people, coming in over New York. This is the first passenger jet to be fully controlled by artificial intelligence techniques. Something goes crazily wrong. The plane banks too steeply. The captain and copilot wrestle with the controls, but the computers seem somehow determined to disobey them. Passengers scream. A suspense sequence in which control is regained only by wrenching out a dozen electronic modules and going to emergency manual control.

VIDEO: The year 2010. A factory in geosynchronous orbit explodes, providing one of the startling media images of the decade: Fiery smoke trails spread out in a perfect star. In the absence of gravity or friction they travel in straight lines for hundreds of miles, like the spines of a giant sea-urchin.

NARRATOR: The most highly budgeted use of artificial intelligence techniques is in the military. The attempt to deploy antimissile weapons in space gave a giant boost to software and hardware development. Extremely elaborate cross-checks were necessary to detect malfunctions and ensure that they could not cause harm.

NARRATOR: (over video) Scientists' experimentation with nature has always provoked fears. The first railway trains were regarded as monsters. Images of Frankenstein's monster reappeared in different forms throughout the twentieth century.

Recombinant DNA gave the ability to create new forms of life, and this more than any other technology gives rise to fears that tinkering with nature may get out of control. Genes can be spliced and respliced to form new genes.

VIDEO: Computer simulations of complex genes.

NARRATOR: Today, biology without gene splicing is as unthinkable as chemistry without chemical reactions, but because it is life that is being changed, there will always be fear and superstition.

> *Preacher*: The Bible says, "Thou shalt not let thy cattle gender with a diverse kind; thou shalt not sow thy field with mingled seed."

NARRATOR: In the 1970s we modified a cell, then we modified microbes, then fruit flies, then a mouse, then dogs. Then in special cases we modified man by creating gene-transplant babies, and furious debate raged about the ethics or wisdom of gene transplants.

New drugs have controlled disease, cured most forms of cancer, reduced mental illness, and extended our life span. There were fears that new viruses might escape from the laboratory, establish themselves in the human gut, and go on multiplying. There were fears that we would plant a slowly ticking biological time bomb.

VIDEO: Newsreel shots of Alfred Vellucci, the mayor of Cambridge, Mass., in 1979, speaking emotionally of breeding monsters in the labs of Harvard and M.I.T., and new diseases escaping to create uncontrollable plagues.

NARRATOR: It has not happened. The precautionary controls are tough. But there may be perils in disturbing the microbial balance, which has been billions of years in the making. We greatly underestimated its complexity. But it is the agricultural products of genetic engineering that now enable us to feed the world's 8 billion people. Without them, untold numbers would starve.

We have no more chance of stopping the scientific mind from probing into the unknown than King Canute had of stopping the tide. But precautions must come first.

VIDEO: The narrator at home.

NARRATOR: The public in general does not want to abandon the comforts technology brings or the expectation that it will bring more and more. The public wants to believe the advertisements.

So in the long run neither the mistakes of technology nor the antitechnology movement had much effect. The evolution of technology raced ahead, driven largely by the marketplace. It raced faster in those areas where the entrepreneur had the greatest freedom.

There was much criticism of capitalism, advertising, and salesmanship. It led to excesses, but overall it adapted the use of technology to the needs of the public. When it went wrong, the marketplace demanded correction. When it caused too much pollution, the marketplace became prepared to pay for cleaner air and rivers. As Churchill said about democracy, "It may not be a good system, but it is the best one we know."

The great problem with the marketplace as a driving force is that it does not look ahead.

VIDEO: Newsfilm of the U.S. gas crisis; 1973 or 1980 gas lines of desperate motorists trying to buy gas. Garage owners carrying guns from fear of attack.

NARRATOR: Action was not taken early enough when the world's demand for energy exceeded its petroleum supply. It took decades to replace petroleum as the predominant energy source.

When petroleum was cheap, there was a lunatic waste of energy. Vehicles used three times the power needed for their journey. American buildings were made too hot in winter, too cold in summer. There were huge expenditures on commuter vehicles that were hardly used except for two brief periods in a day.

NARRATOR: (over video illustration) In the 1980s the waste heat flowing from electricity power stations was the energy equivalent of all gasoline consumed by all the automobiles in the U.S.A. It went mostly into rivers, where it caused pollution, or into large cooling towers. Today it goes into greenhouses or seawater desalination plants or is pumped to heat houses. None is wasted.

The energy crisis of the 1970s and 1980s had multifold solutions but could not be solved quickly. The market-driven technology had failed to anticipate and take action about the problems early enough.

Utterly tragic was the failure to expand the food supply fast enough as the world's population climbed to 8 billion.

VIDEO: Starvation in Africa. Children with potbellies, distorted limbs, and pathetic eyes asking for help.

NARRATOR: Like the energy and mineral crises, the shortages were foreseen well ahead, but strong action was taken only when they reached crisis proportions.

Failure to anticipate, or incorrect belief that current trends will continue, causes economic swings. In industry it causes the business cycle. The economy has always exhibited short-term and long-term fluctuations. These appear today in computer models, which attempt to explain their causes.

VIDEO: Shots of the sea with short, irregular waves.

NARRATOR: A long-term wave exists of much higher amplitude than the business cycle. It is called the Kondratieff cycle. It swings down once every 50 years or so. This long wave is of huge momentum, like a big ocean wave. To stop it or damp it down would require government actions involving far more money than any normally contemplated for economic control. Most economic control actions are directed at the short-term business cycles, which are of much less momentum than the Kondratieff cycle.

VIDEO: Shots of a large swell on a violent sea. Force 8 gale. A boat plunges down a wave far larger than the boat.

ANIMATION: Graphic representation of the Kondratieff cycle.

NARRATOR: The Kondratieff cycle was plunging down during the Great Depression of the 1930s. It was swinging down after 1975. We can trace it back right to the beginnings of the industrial revolution. It has swung upward rapidly since 2000 partly because of the traumatic high-technology reaction to the war. The previous big upswing was from 1950 to 1970.

At each upward surge of the Kondratieff cycle, a new set of technologies is dominant and a major rebuilding of society's infrastructure occurs. The main energy source of the last upswing was petroleum, before that coal, before that wood.

Now there are multiple energy sources: fusion, solar power, and plant alcohol.

VIDEO: A curve follows the long-wave economic cycle from 1720 to 2020 with a series of montage pictures.

NARRATOR: So what is the driving force of this feverish race into technology? It seems to be greed, fear of war, a wish to make money, and a wish to build a better life.

An ocean of technology is rolling in. There are temporary set-backs—depressions, inflations, downswings in the Kondratieff cycle, wars, accidents, and technological disasters—but none of these stop the ocean from rolling in. It will probably continue to do so for many centuries.

Technology constantly appears in new forms, invents new safety controls, solves earlier problems. It is like a great wave. Barriers against it in isolated places do not stop the wave.

The disasters of technology are much publicized. Design feed-back is necessary in all complex systems. This has applied to society as well as to technology. In the long run humankind has usually had the capability to recognize its mistakes and correct them, but often only in the nick of time.

The move to more advanced technologies continues relentlessly. Who determines our technical destiny? Government? Big corpora-tions? In fact, no one entity is in control or can have more than a minor effect. The combined and independent efforts of many people and corporations bring new technology and change society. These people are motivated by the driving forces that have been present since the industrial revolution. Human nature is a constant.

In the long run the driving force is a collective common sense that manifests itself in the marketplace and to a lesser extent in the democratic process. We can influence that collective common sense with education, with television.

The danger of technology and its use being driven by the market-place, or by any short-term forces, is that we may cause long-term problems that are not foreseen. We may cause irreversible trends. We may drift into hazards that cannot be corrected quickly: planetary pollution, damage to the ozone layer, sociological traps, destruction of trees in the third world, inadequate food-growing capacity. We may build systems for war that we will not be able to control when tense confrontations occur.

To avoid causing long-term damage we need to study the future. Detailed forecasts and models are made showing long-term effects of today's actions and trends. To correct the harmful trends or opti-mize the long-term benefits it is necessary that the public understand the long-term effects.

Interactive television is the most powerful communications me-dium in history. It is with television and the press that we harness the collective common sense of the public.

Because politicians and corporations react to their marketplace or constituency in a free country, the best way to steer technology is vigorous public debate. This needs to be both rational and informed.

Professors, gurus, and civil servants work out their views of the future, their ideas about how society's wealth should be spent. The judgment on their views should come from the forum of public debate, like the great forum of Athens. Electronics gives us that forum. Some countries have used it well, others badly. In some, politicians have desperately suppressed the freedom of debate of interactive television.

The best driving force we know is the collective common sense of the public made knowledgeable by inspired leaders and media.

A society that fails to use its television responsibly is asking for trouble.

Installment Six

PAX TECHNICA

PRESENT-DAY AUTHOR: The most frightening and dangerous aspect of future technology is its impact on the machinery of war. The cost per kiloton of explosives has droped precipitously. The world arsenals of nuclear weapons have grown to a size at which they endanger the entire planet. Particularly alarming, the accuracy of missile systems has increased so as to render obsolete or dangerous many of the earlier mechanisms for command and control.

In the past many wars have come into existence for pointless reasons. World War I, for example, started not because any leader overtly wanted the war but as a consequence of the interaction of the military postures and security mechanisms that were established long before the war. An insignificant archduke in Sarajevo was assassinated, and numerous orders were issued for armies to go on alert. One alert caused a corresponding alert. These interlocking alerts created an unstoppable chain reaction of reinforcing events. Everything happened the way it was supposed to according to the rules established years before, but these mechanisms of military response stimulated each other into frenzy. Political leaders lost control of the military momentum that built up. Barbara Tuchman's book *The Guns of August* describes two leaders discussing how World War I started and saying "if only one knew." This story is chilling when translated into the interacting behavior that could occur as nuclear command-and-control mechanisms move from peacetime stability to a situation of mutual alerts, tension, and fear that the other side might strike first. In a potential war situation even precautionary protective actions appear threatening. Once a certain threshold is passed, the players' actions may stimulate each other into escalating alerts with the terrified fear of a first strike by the other side.

The command and control of nuclear forces is a subject that should be of great technical interest to systems people. How does one achieve the level of security needed? How does one prevent war by accident or acts of madmen? Who is authorized to fire weapons under what circumstances? Who can press the button if the president and his close associates are killed? How can the necessary communications links work after nuclear strikes have occurred? Or what happens when there is a communications breakdown? Curiously, I have found it almost impossible to have conversations on these topics with most systems people. Either they do not know anything about the subject or they do not want to think about it. Often they simply assume that somebody must have designed the systems very, very carefully, so we needn't worry.

In fact these are extremely difficult design questions, sometimes without solutions that are 100 percent reliable. The systems can be made safe in peacetime but not so safe when one side thinks the other side might launch, goes into a state of extreme tension, and takes the safety catches off. The dangers become more acute when the missiles acquire pinpoint accuracy so that they can destroy the computer centers and leaders of the other side. The dangers become greater when globe-shattering decisions have to be made in minutes.

Worse, much worse, the immensely complex and expensive command-and-control systems in the United States and the Soviet Union are needed in France, China, India, South Africa, and soon in several third-world countries. The safeguards will be much more difficult to build as nuclear proliferation continues. When in the future an African country or a Latin American country with macho traditions in the military acquires nuclear missiles, can it have either the money or computer expertise or the cool, unemotional checks and balances to build safeguards, and will these safeguards work when it thinks its neighbor might strike first?

Weapons will continue to grow cheaper and much more terrible. It is a global responsibility to design the safeguards. The most urgent subject for history professors ought to be the prevention of future war. To address the subject they would need to learn about computer-ized command and control, telecommunications systems and their failures, the effects of electromagnetic pulses, and expert systems for representing the events that could lead to war. As with systems analysts, it is difficult to find a history professor who is prepared to have a conversation on these subjects.

If we do not build worldwide safeguards, the magnitude of ensu-ing catastrophe is something that no words or television images could do justice to.

VIDEO: The same scene-changing sequence as before is used to slide from the present day to A.D. 2019: a rushing sensation like flying very low at extreme speed through a mountainous valley, but the earth and hills look like endless etched microelectronic circuitry. Electronic music is used, evocative of time travel.

NARRATOR: In 1996 the prime minister of Great Britain made a speech that has haunted the world ever since:

NEWSREEL: British prime minister on television wearing half-moon eyeglasses.

> *Prime Minister:* It is as though the gods were looking down at us, grinning, cackling with amusement. They are playing a game and have taken bets on us. They have given us advanced technol-ogy, and one half has bet that we will destroy ourselves. The other half has bet that we will eventually have the intelligence to build civilizations around the planet with safeguards against destruction. It is a close game. Everyone on earth wants the safe-guards, control mechanisms, and elimination of nuclear weapons. We now understand what the devastation of a nuclear winter

would do to our planet. But national distrust and paranoia win the day. Many brilliant and passionate men have dedicated their lives to the elimination of nuclear weapons, but so far to no avail.

Now the game is heating up. The weapons are spreading. The means of devastation is in the hands of less responsible men. The gods are doubling their bets. Fifty years from now they will know who has won the bet.

NARRATOR: (over newsreel footage) It was surprisingly late in the history of nuclear weapons that the effects of a hydrogen bomb on the earth's biosphere were understood. A hydrogen bomb exploding on a city creates a firestorm thousands of times worse than the one that destroyed Hamburg or Tokyo. The roaring column of fire sends vast quantities of filth into the troposphere.

ANIMATION: The following discussion is illustrated.

NARRATOR: These effects would eliminate sunlight from much of the earth, causing temperatures to drop catastrophically. Much of the earth would drop to below $-15°$ C in a 5000-megaton war and below $-35°$ C in an all-out war. The oceans would be warmer than the land, so violent gales would blow continuously, worsening the effects of the cold temperatures. The term *nuclear winter* was popularized to describe these effects.

The stratosphere is at altitudes over 15 kilometers. This is above the effects of "weather." The stratosphere has no wind and no rain to wash away the particles. The black cloud of filth would remain in the stratosphere, slowly drifting from the areas above the war to the rest of the globe. A severe nuclear winter in the Northern Hemisphere would spread to the Southern Hemisphere. No place on earth would be immune from the effects of a major themonuclear war.

It was 1983 before detailed modeling was done that showed the effects of such a war on the climate and biosphere. The effects were far more devastating than anyone had supposed.

The lack of photosynthesis from the sun would combine with the effects of extreme cold. In some areas the effects would be made worse by the polluted or poisoned atmosphere. When the sunlight eventually returned, it would have contained ultraviolet B radiation at much higher levels than those considered dangerous to plants and animals. The UV-B radiation, caused by damage to the earth's ozone layer, further reduces photosynthesis in plants.

Deprived of food, most animals would starve to death. Their plight would be worsened by radiation from fallout and later by

UV-B radiation, which suppresses the immune system and greatly worsens the effects of rampant diseases. Radioactive isotopes would enter the nutrient cycles, becoming concentrated in the process.

In addition to the black cloud in the stratosphere blanketing the earth, there would be dense smoke at lower altitudes containing acrid and poisonous chemicals, a fallout of radioactive debris over a wide area, and a partial destruction of the ozone layer, causing an increase in biologically dangerous ultraviolet light of several hundred percent. Nucleic acids and proteins, the fundamental molecules of life on earth, would be damaged by ultraviolet light and by the even larger increases of more dangerous shorter wavelengths.

Conifers, which are highly susceptible to radiation, would die over wide areas, leaving highly combustible forests. Many of the fires would burn for months, including fires of coal seams and peat marshes. Much of the hemisphere would be covered with toxic smog from the burning cities, chemical plants, and factories. Roaches, flies, and other insects would survive and multiply rapidly after the death of their predators. After the extreme cold, the surviving rats would multiply rapidly, scavenging on the many corpses.

Plants in the tropics are more susceptible to dark and cold than those in the temperate latitudes. They do not possess the dormancy mechanisms that would enable them to tolerate cold seasons. The majority of plant and animal species in the tropics would have been killed.

Without exception, every American president since the 1970s has expressed a dedication to reducing the world's stockpile of nuclear weapons. Most, starting with President Carter, have stated it is their goal eventually to banish such weapons from the earth completely. Russian leaders made similar statements. But continually, the military on each side has sought superiority over the other side. Weapons technology improved and improved, powerful warheads becoming smaller and cheaper.

If we exclude the cost of the plutonium, the manufacturing cost of 20-kiloton warheads today in countries like Israel and South Africa is less than $20,000 each. Both of these countries have operated breeder reactors and reprocessing facilities and have almost certainly stockpiled plutonium.

Technology improvement, mutual distrust, and the power of the military always seemed to outweigh the advances gained with painfully slow negotiation and posturing in the disarmament talks.

VIDEO: Film and graphics combine to illustrate the systems in the following discussion.

NARRATOR: Vast sums were spent on research into defensive systems, intended to destroy missiles shortly after launch. These "star wars" systems were immensely expensive. They steadily evolved from impractical visions to working systems. They required artificial intelligence computers that could process billions of inferences per second. They required lasers and particle-beam systems that were themselves powered by nuclear devices.

There was a hope that defensive weapons would become good enough to neutralize offensive weapons. The vast military budgets were acquired on the premise that the earth would be freed from the threat of missile attack.

However, numerous steps were taken to protect the offensive weapons from the systems in space. Space beam systems were built to attack space beam systems. Elaborate decoys were built. Bombs were built that would release intense electromagnetic pulses (EMP) that would disable computer circuits, chips, and telecommunications systems over a large area.

As the decoys released by intercontinental missiles became more sophisticated and more numerous, the necessity of destroying the missile during its boost phase became more vital. Only space-based weapons could achieve this. The boost phase lasted several minutes when the first space-based defense systems were designed. In response to new defenses, the boosters were redesigned so that the boost phase lasted only 80 seconds. Then much faster-acting systems had to be designed for attacking the boosters. Such systems needed split-second decisions, which could only be made by computers. It was increasingly necessary to rely on software of extreme complexity, pushing the state of the art, on systems that could be tested only in a simulated fashion. Putting cataclysmic split-second decisions into the hands of such software was highly controversial.

The complexity of the attack, counterattack, counter-counterattack, and counter-counter-counterattack systems became so great that no computer could predict with assurance that outcome of a war that released this nuclear-electronic frenzy. The war games computed probabilities, but the uncertainty factors remained great. A policy of mutually assured destruction (MAD) was replaced by a policy of mutually assured uncertainty (MAU). This applied equally to both sides (MAU-MAU).

It was clear that many missiles would not get through the space-based counterattack systems, but some would. So each side felt the need to increase its chances by building more missiles. Mass-production factories in Russia produced more and more nuclear warheads so that *some* would penetrate the star-wars screen. The United States did the same so as not to be left behind the Russians.

VIDEO: A mass-production line with robots handling dangerous materials for the manufacture of nuclear weapons.

NARRATOR: By the perverse arguments of nuclear war, the existence of *defensive* systems caused more *offensive* weapons to be built.

Much larger rockets came into use because of the needs of the space program spurred by Russia's attempt to visit Mars for the centennial of the Russian revolution. With larger missiles and more compact bombs, one missile could carry 80 MIRVed warheads, so the warheads were mass-produced.

The term *gigaton* replaced *megaton* to describe world arsenals. (A gigaton is equivalent to a billion tons of TNT). The world today has over 50 gigatons of nuclear weapons. That is over 6 tons of TNT for each human being.

The threshold that could cause a nuclear winter is less than half a gigaton. (Even 100 megatons could cause climatic disaster if released on 1000 cities.) We are adding almost a gigaton per year to the world's supply of such weapons.

VIDEO: A chart shows three piles of plastic counters labeled "U.S.A.," "U.S.S.R.," and "Other." Counters labeled ½ gigaton are being added to each of the three piles. (See Fig. 6.1.)

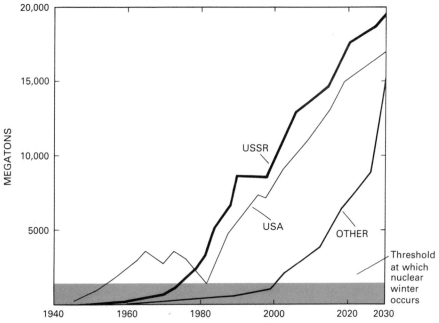

Figure 6.1   Megatonnage of weapons in the nuclear arms race.

NARRATOR: Today a nuclear winter could be much worse than it might have been in the 1980s, when it was first discussed. The world arsenal of nuclear weapons has gone up by over 40 gigatons, and today many of the targets would be in the Southern as well as the Northern Hemisphere. The entire earth would be blanketed with the life-killing darkness. The ice would be everywhere.

> *ANIMATION*: Bringing Fig. 6.2 to life. Projected designs of survival abodes.

NARRATOR: If one third of the Russian and American nuclear weapons had been used in 1985, the earth's average surface temperature would have dropped to −15° C for about several weeks. If one third were exploded today, the average temperature would drop to −35° C

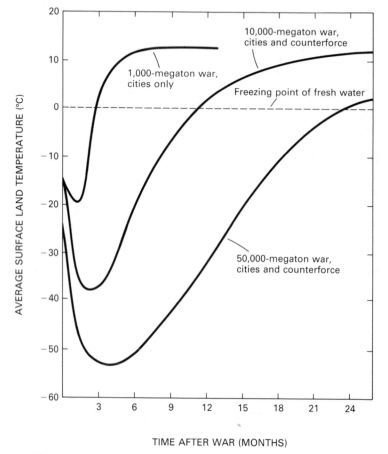

Figure 6.2   The average temperature of land areas after nuclear wars in both hemispheres.

and stay below freezing point for about a year. The plant and animal death on the surface of the planet would be almost total.

The only plants and animals that would survive would be those carefully protected in specially designed nurseries and Noah's Ark projects.

Zoos exist deep underground, in tunnels in the Swiss Alps, in caves in Chile, in electrogenerating stations in the rock of New Zealand.

VIDEO: The camera shows the underground zoos with animals lit by floodlights.

NARRATOR: The Woomella game park, 100 miles north of Perth, is enclosed in a vast plastic enclosure. Its lights and pumps are powered by many 60-kilowatt windmills, which would turn furiously in the coastal gales of a nuclear winter. The water holes are fed from deep wells by windmill-powered pumps. In addition to its animals, the enclosure contains vast nurseries of plants, including all the exotic species of flowers that inhabit Western Australia. The park is a major tourist attraction, but the fate of the animals in a nuclear winter is open to question if people around are starving.

VIDEO: A montage of the animals, enclosed parkland, water holes, windmills, and unique flowers of Western Australia.

NARRATOR: The nuclear winter effects have been studied in great detail for four decades. It was feared that humanity itself might be exterminated. Now that will not happen even if the worst war occurs because of the specially equipped camps and shelters. But when the sunlight returns, there may be only a few million human survivors, and vast numbers of species would be gone forever.

The dangers of a nuclear winter suddenly seemed much more alarming when it was realized that third-world nations were growing their own nuclear arsenals. It was not only the United States or the Soviet Union that could cause devastation to the biosphere.

ANIMATION: A world globe rotating with graphics of mushroom clouds showing many countries with nuclear weapons.

NARRATOR: It was clear that before long a war in India, China, Africa, or South America could cause a climatic disruption and radioactive fallout that would play havoc with the lives of people in innocent nations. The total megatonnage of the third world was growing almost as fast as that in the U.S.A. or the U.S.S.R. had 40 years earlier.

As third-world countries acquired nuclear weapons and delivery systems, they needed intricate computerized launch control facilities, early warning networks, and command-and-control systems. Like the United States and Russia, they developed elaborate nuclear strategies and preprogrammed lists of targets. In great secrecy they developed their equivalent of the U.S. SIOP (Single Integrated Operational Plan).

Without exception, the objective of third-world military commanders was to achieve a first strike, not a retaliatory strike, if nuclear war came. The first strike was designed as far as possible to destroy the military command system of the enemy nation and to prevent it from responding.

VIDEO: A belligerent top general in Argentina being interviewed.

> *General*: Of course we have a first-strike policy. Any other policy is suicide. You've got to compare going first with not going at all. Every nuclear nation has a first-strike target list preprogrammed. The only country that pretends it does not is the United States.

NARRATOR: Nuclear strategy changed drastically without the public realizing it when a new generation of missiles of unprecedented accuracy came into service in the late 1980s. They could land MIRVed hydrogen bombs within 100 meters of their targets.

The movie industry was still portraying the Cheyenne Mountain Complex as the center of nuclear war decision making at a time when the new missiles had rendered it obsolete. The famous command-and-control center built into the granite of the Rocky Mountains near Colorado Springs had 87 large computers. Its main command post had the wailing sirens and giant display screens depicted in many movies. In its day it was the apex of NORAD Command and Control, where the NORAD commander in chief made his formal assessment of the state of the war and had instant telecommunications links to the U.S. president, the Joint Chiefs of Staff, the commander in chief of SAC, and other top officials.

By the 1990s one MIRVed missile from a Russian submarine, with a 10-minute flight time, could destroy or totally isolate the Cheyenne Mountain Complex.

Deep shelters of any type are no longer safe if the enemy knows their position—and it can be assumed that they always do. A direct hit with a hydrogen bomb will destroy or totally seal off any shelter, no matter how deep. There is, therefore, a good chance of destroying the top officials of a country, the military commander in chief, and much of his chain of command.

In principle, the military and government leaders could take off in a plane to avoid being killed, but in practice the 10-minute missile trajectory may leave them no time to do so.

The ability to destroy deep shelters with certainty and to destroy the leaders of a country and their computers created a situation like that between two gunfighters in the old Wild West.

VIDEO: A scene from an old western movie with two gunmen taunting each other in a bar until both reach for their guns and fire.

NARRATOR: When two antagonists have first-strike policies in a period of tension, they need hair triggers so that the strike can occur very fast if the situation reaches that point.

Hair triggers on the vastly complex nuclear machinery may be made safe in peacetime, but in periods of tension, when one side thinks the other might act, they become very dangerous.

NEWSREEL: The British prime minister with half-moon eyeglasses, 1996.

Prime Minister: The climatic aftereffects of a hydrogen-bomb war will be devastating to nations not involved in the war. An uninvolved nation may be totally destroyed by the nuclear winter and radiation effects. It is absolutely unacceptable to all nations of the planet that such a war might happen by accident or because of squabbling among nations that constitute a minority of the planet.

NARRATOR: When it is possible to destroy quickly the top leaders of a nation, the question arises, Who should have their finger on the nuclear button? The pinpoint accuracy of today's missiles has caused nuclear nations to target the leaders of antagonist nations.

ANIMATION: A hierarchical chart showing a national command organization. The upper blocks are deleted one by one with simulated explosions.

NARRATOR: In the United States the public often thinks that only the president has the ability to authorize the firing of nuclear weapons. The president is accompanied everywhere by a member of the White House Military Office, who carries a box referred to as "the football" because of its shape.

VIDEO: Newsreel shots showing "the football."

NARRATOR: The football contains the GO codes and instructions such that the president can, in effect, press the nuclear button.

In reality this view of how the U.S.A. might enter a nuclear war has been little more than folklore. The public would like the head of state to be the only person who can authorize the use of nuclear weapons. They would like the weapons to be secured with a secret code that only the head of the state can release. This, however, is not practical in the United States or any other country. Authorities on nuclear strategy emphasize that if a weapon can be smuggled undetected into a nation's capital, the detonation of this should precede a full-scale launch. The head of state and those immediately around him may be the first to be killed if war happens. No country could run the risk of its GO codes being lost in the critical seconds.

NEWSREEL: The Argentine general being interviewed.

> *General*: There is no way we will risk having to search around in the rubble in order to find the key to fire the missiles.

NARRATOR: This dilemma might be dealt with by devolution of authority. The president's office automatically passes to the vice president, the speaker of the House, and to others in a sequence specified by law. After a first strike, a low-level minister could suddenly find himself commander in chief. In a war situation it might be difficult to determine that the secretary of education has suddenly become commander in chief. In any case there may be justified reluctance to give the secretary of education access to the GO codes. Civilian successors to the president, even in the United States, may be almost completely unacquainted with the command-and-control complexities of the nuclear systems. In India, China, and Iran the situation is much worse.

> *NEWSREEL*: Bellicose posturing of an Idi Amin–like head of state with a vast array of medals.

NARRATOR: There was particular concern about which third-world fingers could be on a nuclear button after one particularly belligerent Moslem head of state claimed that God had given us nuclear weapons to cleanse the earth.

> *VIDEO*: Early warning radars. A computerized command-and-control center. Missiles rising from their pods.

NARRATOR: The expensive machinery of nuclear response may be suddenly useless if the country is like a chicken with its head cut off. The best first-strike policy is to cut that head off before the response machinery comes into action.

Because of this danger, constitutional devolution of authority is entirely unacceptable to the military. The military wants to be able to fire its weapons if the head of state and his close associates are gone.

VIDEO: An African military officer responding in an interview.

Officer: There's an informal gentleman's agreement about how things will be done. There has to be. If war happens, the military commanders can't wait for hours to find out who now officially has authority. In missile warfare, every second is vital.

Dissolve to scenes of a submarine with nuclear weapons on the ocean bed and the cramped life of the black sailors on board.

NARRATOR: The weapons most difficult to locate are those on nuclear submarines, which can lurk near the ocean bed undetected. These may be used either as first-strike weapons or as the ultimate retaliatory weapon. A large nuclear submarine today has 48 missiles each with 25 MIRVed warheads. One submarine may have 1200 hydrogen-bomb targets. Some third-world countries have several weapons-carrying submarines. These alone have the capacity to start a nuclear winter.

Unfortunately, it is difficult to communicate with submarines on the ocean bed.

If all else is destroyed, the nuclear-missile submarines can still fire their weapons. In the U.S.A. and elsewhere, the submarine commander can fire without a directive from the president or his delegated representative. In a high-level alert the submarine missiles can be fired without GO codes. Most radio and other signals cannot reach the depths of the sea, so it may be difficult for submarine commanders to know exactly what has happened in the mayhem of nuclear war.

The firing systems on a submarine have dual key systems and are thoroughly secure. No deranged individual could fire the missiles alone. Nevertheless, in a severe alert, the submarine commander in conjunction with others on board is able to launch the missiles without external orders. In the design of submarine missile orders, "fail-safe" means that if the mainland systems are destroyed, the submarine missiles can still be fired.

VIDEO: A red-alert situation in an Indian nuclear command center. Sirens blaring. Lights flashing. Terrified-looking Indian staff rushing to their positions. Planes taking off.

NARRATOR: The first target in any nuclear exchange is likely to be the nuclear command center. Most countries therefore have airborne command centers permanently in the air. In Brazil converted Boeing 737s are used, with no windows other than in the cockpit. These planes are filled with elaborate communications equipment. Every eight hours such a plane takes off with an air force general and battle staff on board, relieving a similar crew in a similar plane. In principle the flight plans are secret. In practice these planes are always tracked by the air traffic control satellites.

If a surprise attack destroys the command center on earth, the general in the airborne command plane would be likely to become the "doomsday officer." He has the ability to transmit GO codes to missile-launch sites, planes, and submarines.

> *VIDEO*: A scarlet box with three locks on board the airborne command plane is opened by three airforce officers with separate keys. When opened, it makes a loud clacking noise. A general removes papers from inside the box.
>
> > *General*: These are the authentication codes. With these we can compose and transmit the emergency orders to fire.

Shot of missiles rising from their silo.

NARRATOR: In a war situation the general on the airborne command plane is only likely to survive himself if the enemy is neutralized. He would be motivated to order the firing of every available weapon. In fact, he may have no ability to know what is really happening. EMPs, enemy jamming, or destruction of transmitters may put him out of touch. But he is still likely to decide to fire everything.

VLF (very low frequency) transmission can survive most attempts at enemy jamming. However, it requires a vast antenna.

> *VIDEO*: The following discussion is illustrated with film clips.

NARRATOR: The airborne command planes in some third-world countries copied a U.S. system in which an antenna wire 5 miles long was uncoiled behind the plane, its end attached to a small drogue. This antenna weighs almost a ton and makes the plane very difficult to fly. The plane can make only very gradual turns. In storms or turbulence the antenna sometimes starts to whiplash. Sometimes the antenna breaks off or the crew has to release it because it makes the flight too unstable.

As with nuclear-missile submarines, the generals in airborne command planes have the ability to issue launch orders on their own in certain circumstances.

No nuclear weapons have been launched by malevolent individuals in the U.S.A. or the U.S.S.R. because of a highly intricate web of controls—controls that must still work if the president and his successors are killed, if part of the national command is destroyed, if computers fail, or if telecommunications are disrupted. In the 1980s and 1990s, when other countries were setting up nuclear command structures, the United States and Russia set up highly secret teams to advise new nuclear nations on how to avoid war by accident. It became immediately apparent that the intricate safeguards of the old nuclear powers were not being used fully. A junior officer certainly could not launch an unauthorized attack, but a general might be able to. The lower-level safeguards were fine; the highest-level safeguards were not.

The problem was made worse by the fact that many of the new nuclear nations had frequent *coups d'état*. The head of state went to bed frightened that he could be deposed by morning and his family killed. As the best time to attack a country is during the immediate aftermath of a palace revolution, how can an intricate web of nuclear safeguards be made to work during such a period?

A U.S. or Russian head of state would be extremely careful to avoid a drift toward nuclear hostilities, but the same confidence was not felt about some third-world leaders. Nuclear weapons are possessed by two African dictators, both of whom have practiced torture and murder on a grand scale and who brag, at least in private, about the nuclear havoc they will wreak if their enemies are not careful.

It is highly unlikely that the U.S.A., Russia, or Europe would have a nuclear war with each other by accident or because of an undesirable political drift. In the third world, however, it seemed that anything *might* happen. Their nuclear systems were designed with hair-trigger mechanisms. It is unlikely that they have the expensive and comprehensive safeguards that exist on the war machinery of major powers. Highly unstable leaders come to power, including some whom historians will describe as madmen. Because of the vulnerability of command-and-control centers, command of nuclear forces is highly distributed, and often it is not clear who *could* have a finger on a nuclear button. Both airborne and submarine commanders have the capability to *launch under duress*. On several occasions the 5-mile-long VLF antenna on the Brazilian 737 airborne command centers has fallen off in flight. There appear to have been many nuclear false alarms in the third world, and often a tense confrontation that causes unintended crisis interactions results. One side can take steps without knowing that they are highly provocative to the machinery of the other side. Confrontations in the third world appear much more volatile than in the major powers, and once a crisis turns into a strategic alert, all safety catches are taken off.

NEWSREEL: The British prime minister in 1996.

> *Prime Minister*: No nation can tolerate the possibility that other nations play irresponsibly with nuclear fire. Given human history, it is inevitable that national confrontations will occur that will put the systems for nuclear first strike into a level of tension and terror in which an incident will occur. One incident may trigger the firing of arsenals of hydrogen bombs, causing utter wreckage of the planet's biosphere.

No society tolerates arsonists or saboteurs. Society needs a police force that prevents violent criminals from running amuck. Today, like it or not, we have a world society. The world is instantly linked via telecommunications. Worldwide nations are entirely dependent on their interchange of food and minerals and manufactured goods. One nation starting a nuclear war can destroy the environment for us all.

A world society must have a world police force. That police force must have the power to prevent criminally irresponsible violence. In the long run, the only way civilization on earth will survive is to have a police force that prevents the possession and use of nuclear weapons.

We need the police force today. The weapons are spreading in the third world as fast as they spread 40 years ago in the United States and the Soviet Union. Unlike 40 years ago, the weapons now have pinpoint accuracy such that a first strike is the dominant nuclear strategy. This causes tensions that the command-and-control systems can barely handle. And the third-world command-and-control systems lack the safeguards that have grown up in the major powers.

Eventually the United Nations ought to be the world anti-nuclear police force. But today we have to face the reality that it does not have the power or the budget. In this situation it is essential that the strongest members of the planet act to preserve the safety of the planet. It is the moral duty of the United States, the Soviet Union, and Europe to act to prevent planetary destruction.

The military of all nations have always preferred to spend their vast budgets on weapons of attack rather than weapons of defense. Bombs and guns are more macho than shields and handcuffs. You get medals for attacking, not for defense. Today the consequences of launching the weapons of attack are so devastating that we desperately need to spend all of our technological brilliance on defense.

The prototypes of X-ray lasers and other beam weapons are now working well enough for us to plan to deploy them on a large scale. Instead of Russia and America building these weapons to zap each other's missiles, they should be building weapons to jointly police the entire planet.

I hereby call upon the nations of the first world and the Soviet bloc to join forces in protecting the planet. We must eliminate plutonium worldwide and place the severest sanctions on nations that will not cooperate in this. We must combine intelligence and surveillance data about the world's missile installations and nuclear delivery systems. We must combine in the use of beam weapons to prevent hostile missile launches anywhere on the planet. If third-world nations know that their missiles will be destroyed when launched, they will not build missiles. If they know that their bombs cannot be delivered, they will not build bombs. If they know that the sanctions against them can be devastating, they will cooperate in making the planet safe.

It will take decades to build a world police force for protecting the planet. It is essential to start now. That police force needs advanced technology. Its capabilities will improve with time.

If we do not combine to use our technology for protection, the gods that have placed their bets on our destroying ourselves will win their bets.

NARRATOR: This historic speech got much publicity. The image of the gods playing a crap game with man and technology offended many but haunted everyone. But no action was taken. The British prime minister repeated that if nothing was done, the weapons would indeed be used.

Then, suddenly, the 1999 war occurred, and for a day the world was utterly terrified.

This war did not destroy the biosphere. It was too small and localized. But it severely affected the earth's climate for a year, spread devastating radioactivity, and people are still dying today from radiation-induced cancer. The aftereffects of the bombs were measured in great detail. The measurements improved the precision of the nuclear winter models. With a growing sense of horror the public realized what would have happened if the war had spread. The measurements were fed into computer models, which showed that the total devastation of the planet would result from using even a tenth of the available nuclear arsenals.

The average surface land temperature of the Northern Hemisphere would have dropped to below $-20°$ C and remained below $-10°$ C for a year. The cold and darkness would have killed most types of plants, especially as the nuclear winter would have begun at the start of the growing season.

Three feet or more of ice would have formed on most bodies

of fresh water in the Northern Hemisphere. Photosynthesis would have ceased in algae, which form the basis of most marine food chains. Reproduction of these plants (phytoplankton) would have stopped. This, combined with pollution from the toxic and radioactive runoff, would have resulted in the death of much aquatic life.

There would have been a few scattered, weak human survivors from such a nuclear winter, their social and value systems utterly shattered. Untold numbers of plant and animal species would have been lost forever. The biosphere would never have recovered to today's state.

> *VIDEO*: A montage of images of the effects of a nuclear winter. Darkness and ice. Dead cattle. Billowing clouds of black pollution. Dead fish in rivers. Fires in forests of conifers visible across fouled marshland. Dead crops. Dissolve to a cheetah racing through long grass.

NARRATOR: For two decades after the 1999 war, technology for preventing nuclear confrontation was deployed with a furious energy. Russia and America both deployed elaborate surveillance satellites, X-ray lasers, and other beam weapons. Ultra-high-speed computers processed the surveillance data and kept constant watch for missile launches. Each side was careful to notify the other of its intended launches.

> *VIDEO*: A large X-ray laser beam weapon in orbit. Large-screen consoles for defense monitoring. Computer-controlled mirrors in space for reflecting laser beams. The camera zooms slowly down a vast chemical laser. Animated illustration of a submarine-launched missile being destroyed in flight.

NARRATOR: The antagonism between gunfighters in an old western movie can change to respectful cooperation if they realize that they are both threatened by some external force. The beginnings of cooperation between the major nuclear powers came when it was clear that third-world nations were building enough nuclear firepower to cause climatic devastation.

Much had been written about the spread of nuclear weapons to third-world countries, but after the 1999 war it suddenly seemed a desperately urgent subject. At that time 18 countries were thought to have such weapons.

America and Russia watched each other like prowling animals but cooperated in their surveillance of the third world. Both deployed beam weapons that would destroy hostile missiles if they were launched by other nations. Intelligence data about third-world nuclear activities were exchanged. Economic incentives, sanctions, and threats were used to have breeder reactors and reprocessing facilities that created plutonium dismantled, replaced with other energy sources, or temporarily operated with international security forces.

Various smaller nations protested about the large "bullying" nations, but most abandoned the attempt to stockpile hydrogen bombs and build sophisticated delivery systems. The major powers made it clear that such weapons would be blasted with X-ray lasers, so there was no point in spending the vast amounts of money they required. Small wars flashed around the globe like summer lightning, but the technology and policing of the major powers made the dangers of nuclear war seem to recede.

The technology-enforced nuclear policing by the major powers became known as *Pax Technica*. Without it the third-world growth of nuclear forces would almost certainly have been uncontrollable. Beam weapons were highly controversial when first proposed for use against Russia, but they were to lead to a technology of defense without which nuclear proliferation would have become immensely dangerous. The British prime minister famous for the "gods looking down" speech has made several speeches since retirement, commenting on the state of play of the gods' game.

NEWSREEL: The former prime minister, still with half-moon eyeglasses, 2014.

Ex-Prime Minister: It is now 15 years since our second war with nuclear weapons. Today it looks less likely that there will be a third such war. Pax Technica, like Pax Romana two thousand years ago and Pax Britannica in the nineteenth century, may preserve nuclear peace for the time being. But no one should be allowed to become complacent about the terrible dangers that will always be present until we reach saner times when our bombs are dismantled.

Looking back, many historians say that the war of 1999 was inevitable. It seems as though we had to use our hydrogen bombs once in order to take seriously the full horror of such a war. Now we live with the detailed images of it on our wall screens. Let us hope that the same devil's logic does not apply to the nuclear winter—that we have to do irreparable damage to the planet before our governments can really take it seriously.

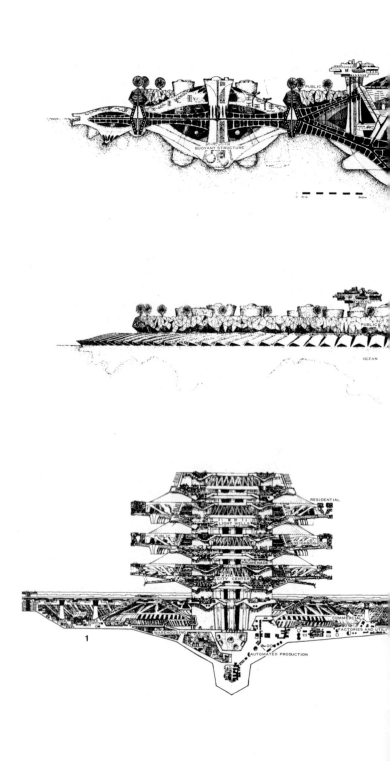

1

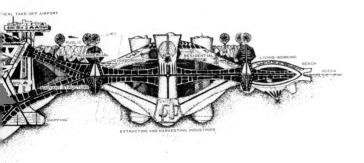

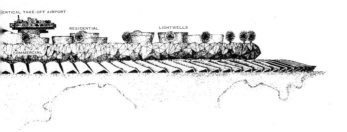

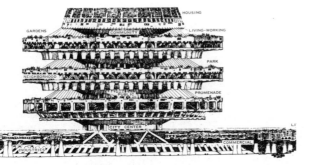

2

Installment Seven

ONLY ONE EARTH

PRESENT-DAY AUTHOR: In the nineteenth century land to grow food was abundant. In the mid-twentieth century population exceeded the capacity of the land to feed it without technology such as tractors, artificial fertilizers, and transportation. Without realizing it, humankind had entered a technology trap. It had become dependent on technology.

In 1960 the world seemed overpopulated, but by the year 2000 it will have twice as many people, and in spite of all the efforts with birth control, the third-world population will be growing faster than ever. Since the 1960s a back-to-nature movement has been popular. However, the abandonment of technology would mean mass starvation.

We will not be able to feed the population of the near future with conventional agriculture and fuel sources. So we need more exotic technology: hydroponics, artificial food, new fuel sources, genetic engineering. The technology trap is deepening.

Technology has caused some of the world's problems, but the only way out of them is more technology. This time, however, we understand the need to avoid uses of technology that are ecologically destructive.

By the year 2020 the planet will be bulging at its seams—too many people, too little food, too much pollution, too fast a growth rate, and, most stressful of all, 5 billion third-world people wanting to have cars, to build factories, to use more energy, and to consume resources like the rich countries. The stresses will greatly increase the probability of war. Since World War II there have been nearly 100 outbreaks of war. All but a handful of these have been between relatively poor nations. As nations become affluent, the likelihood of them going to war becomes much lower. When the affluent nations trade, show each other's television, and flood each other with tourists, the probability of war is lower still. There is a very strong statistical correlation between national poverty and war. It is almost inconceivable to imagine a war between Britain and Germany again.

Among professional people the world is remarkably homogeneous. Computer experts traveling the world find other computer experts with whom they instantly communicate. They have the same problems of making technology work and share the common language of technology. An American IBM emloyee finds that he has more in common with a Japanese or Brazilian IBM employee than he has with some people at cocktail parties near home. Ballet dancers, filmmakers, accountants, and other professionals who travel worldwide find a similar rapport.

In a century or more, the affluence and understanding brought by technology will probably have spread throughout the whole planet. Different religions, one hopes, will be prepared to live in peace with one another as do the very different religions in Japan and America. It is difficult to imagine an Ayatollah Khomeini coming to power in countries with an affluent public and numerous television networks. Imagine what Ted Koppell would do to the Ayatollah. Combined with the spread of affluence, television, and tourism will be the fear of weapons systems becoming ever more terrible and the understanding that great caution is needed to avoid their use. The mechanisms for policing, military intelligence, and peace-keeping will become highly efficient.

The dangers lie in the shorter term. How do we evolve from today's world to a world of greater equality, education, and meaningful life styles? To a world where technologies that consume excessive resources are avoided? To a world where population growth has been controlled?

With jumbo jets, satellites, and computer networks the world is a small place. How do we achieve the understanding and trust among nations fast enough to defuse the chance of wars with future weapons?

VIDEO: The same scene-changing sequence as before is used to slide from the present to A.D. 2019: a rushing sensation like flying very low at extreme speed through a mountainous valley, but the earth and hills look like endless etched microelectronics circuitry. Electronic music is used, evocative of time travel.

VIDEO: Slow zoom through the stars of the galaxy. Dissolve to a beautiful picture of the planet Earth.

NARRATOR: Science fiction tells stories of voyaging easily to other civilizations. Science and technology, on the other hand, are increasingly telling us that we are *alone*, isolated by unimaginable distances. There will be no such visits at least in the foreseeable future. With the fastest spaceships of today it would take a billion years or more to reach the nearest inhabited planet, even assuming that we could find it.

Technology can work miracles, but one miracle not available to us is to travel to other planetary systems. We are alone.

Our nearby planets are extremely inhospitable.

NEWSREEL: Footage of the Russian landing on Mars in 2017 showing the desolation of the landscape.

NARRATOR: The rocks of Mars, the Hell of Venus, and the maelstrom of Jupiter make us appreciate the beauty of Earth. We will not make other planets hospitable, at least not for thousands of years. There is no substitute for the beauty of Earth.

VIDEO: Images of barren planetary landscapes. Dissolve to a picture of Earth small in the center of a black screen.

NARRATOR: We are alone and isolated. For 80 years we have monitored the radio frequencies with powerful computers searching for a trace of signals that would reveal intelligent life elsewhere. There are no such signals. The dreams of science fiction remain only dreams.

VIDEO: The camera pans slowly through an image of the night sky with music that suggests isolation. Dissolve to the picture of Earth, again, small in the black void.

NARRATOR: Our planet is incredibly beautiful.

VIDEO: A lyrical sequence showing the beauty of Earth's oceans, streams, and woods. Sunlight beams on a misty morning through the leaves of oak trees round a dark lake. Kingfishers dart by the glittering water.

NARRATOR: Earth, however, is far more fragile than was originally thought.

VIDEO: We see a sequence of severe ecological destruction. Factory pollution. Smog over Los Angeles. A forest of dead trees. A major oil slick: deep sludge where once there was sand. News film of a South Carolina beach with seabirds black and covered with oil. A Geiger counter ticking furiously.

NARRATOR: For the first two centuries of industrial growth, the earth's resources seemed unlimited. There were new lands to conquer, new pastures to till. Nature absorbed our effluent.

VIDEO: Scenes of nineteenth-century treks. Wagon trains. Horse riders galloping across the plains.

NARRATOR: There is always the possibility of technological or scientific mistakes. As technology becomes grander in its scope, the severity of mistakes becomes more serious. If a catastrophe does occur, we analyze it in the greatest possible detail and try to make sure that it can never happen again.

VIDEO: Images of the Midwest dust bowl; a sign saying that shellfish eating is prohibited; a hand reaching to the dashboard of a car and pressing a button labeled "Oxygen."

NARRATOR: But in the latter half of the twentieth century, there was a real risk of the earth being damaged, not by a technological mistake but by an attitude of mind that treated the planet as an open system to be spoiled and wasted without thought—because there was always more. A pernicious attitude, which we are at last beginning to change.

VIDEO: We see Aborigines hunting, and some of their culture, so very much at one with the environment.

NARRATOR: Once humans cease to be hunter-gatherers, interference with the environment starts. Humans chop trees and kill creatures that attack their crops.

VIDEO: Scenes of England from the air—Cranborne Chase, a mass of suburban houses and factories.

NARRATOR: This part of England was once thick forest with as complicated an ecology as the African jungle.

NARRATOR: In the twentieth century the pace increased at an enormous rate: soil erosion; the cutting of trees in India and Africa, which led to dust bowls; monoculture with fertilizers, which weakened the soil even of the abundantly fertile American Midwest; strip mining; the dumping of poisonous chemicals in lakes.

Fertile land was lost to roads and buildings at the rate of over 300,000 acres a *week*. While the world population was growing fast, the land that was growing food decreased.

VIDEO: A sequence of human destruction of nature: faster and faster destruction. Huge areas of forest felled with devastating efficiency. Fields turned into mud and builders' rubble. Industrial effluent belching into a river. Vast "jungle crusher" machines in Brazil.

NARRATOR: A understanding that there were limits to growth dawned in the 1970s. A computer model was produced at M.I.T. for the Club of Rome, which today looks as simplistic as Galileo's first telescope.

The behavior of any complex system is determined by the relationships of its components. A change in one part of a system affects other parts, often with a time delay. These in turn affect other parts, and so on.

The M.I.T. model treated the world as a single complex system. It represented the relationships among components such as minerals, energy supply, food supply, population, and pollution. It used mathematical equations to show relationships among economic, technical, political, psychological, and other forces.

VIDEO: Figure 7.1 is revealed progressively from left to right.

NARRATOR: Land for growing food declines. Pollution increases. Eventually the food per person and industrial output per person fall severely.

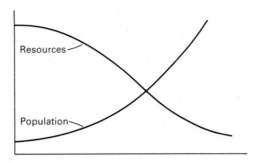

Figure 7.1

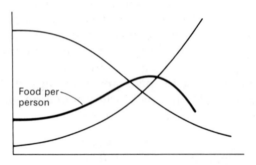

Figure 7.2

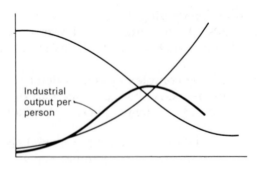

Figure 7.3

VIDEO: The curve in Fig. 7.2 and then the one in Fig. 7.3 are added from left to right.

NARRATOR: Most obvious attempts to solve the problems made the situation worse by permitting faster growth, which eventually strains the resources still further. Only by slowing growth could catastrophes be prevented in the future. All this was traumatic for Western society, where every nation had been operating for many years on the assumption that its GNP would grow by a certain sizable percentage every year: There would be more goods and services, more money, and so, for everyone, a little more of everything.

The limits-to-growth model said that we were running out of energy, running out of raw materials, and not able to grow enough food for the burgeoning world population. Worse, there were lengthy feedback delays in the relationships among the variables, so the disastrous effects would not make themselves felt until the growth had gone too far. There would be desperate shortages, cutbacks, and famines.

ANIMATION: The following discussion is illustrated graphically.

NARRATOR: A system that possesses the three characteristics of rapid growth, feedback delays, and environmental limits is inherently unstable. Because the rapid growth persists while the feedback signals that oppose it are delayed, the physical system can temporarily expand well beyond its ultimately sustainable limits. It overshoots, and there are then painful corrections. During the overshoot the environmental carrying capacity may be so diminished that it can support only a much smaller population or a lower material standard of living.

Long-range doomsday predictions have always contained a flaw, at least so far. They underestimate human ingenuity in solving the problem. As soon as the dangers are recognized to be genuine—and they are usually recognized too late for stability—we set to work to avert the danger.

The problems that become catastrophic in the limits-to-growth model are that as the population grows there are declining energy sources, declining raw materials, insufficient food, and excessive pollution.

VIDEO: Montage of drilling rigs, giant shimmering refineries, oil wells in the desert.

NARRATOR: Energy crises recurred periodically as petroleum became more scarce and more expensive. It was a problem that took decades to solve, but it was attacked from many different directions. Synthetic petroleum was created. Plants were grown that produced alcohol rather than other forms of fuel. After half a century of expensive research, fusion power stations came into operation.

VIDEO: A montage of solar panels, giant fusion power stations, a power-generating satellite in geosynchronous orbit, two-bladed windmills, a California wind farm, gasohol, plants grown to produce alcohol, a sea farm harvesting new forms of energy-producing seaweed.

NARRATOR: While major industries grew up to build new energy sources, a most important factor was the conservation of energy. Microcomputers were used everywhere to make facilities energy-efficient. Cars operate on a quarter of the fuel they used in the 1960s. Microcomputers open and close blinds and heat storage ducts. Machines were miniaturized, digitized, and automatically controlled to save energy. The era of crude, fume-belching energy wastage was a temporary aberration in history.

VIDEO: An overloaded truck in 1950 struggling uphill, pouring out smoke. A massive traffic jam in São Paulo, honking horns, and a purple haze of pollution. A roaring flame on a stack in a refinery. A 1955 DeSoto with a large chrome grille.

NARRATOR: As raw materials became used up, we humans were equally ingenious in finding replacements.

VIDEO: A Victorian mill town with a forest of factory chimneys belching smoke into a sky deep red at sunset—dissolve to spotless laboratory-like factory with computers, and workers in laboratory coats.

NARRATOR: Skyscrapers were no longer made with steel skeletons but with wood and concrete, built up in stages floor by floor. Fiberglass replaced metals. Optical fibers replaced copper telephone cables.

The first optical fibers carried 1000 times the information of copper cables. Today's optical fibers carry 1000 times the information of the early fibers.

Computers using a lot of power were replaced by chips using almost no power. The chips and optical fibers were made with silicon, the world's most common mineral, the sands of the desert.

VIDEO: Montage of rolling desert sand dunes, a red-hot crucible, images of complex logic on chips, optical fibers.

NARRATOR: New alloys were created in space, where the near-perfect vacuum and absence of gravity enabled giant crystals to form and new alloys to be created that would separate out if mixed in gravity. The ultrapurity of space allowed the manufacture of new materials in which a trace of impurity would cause harm.

VIDEO: Manufacturing facilities in a space station with hundreds of spider-like robots.

NARRATOR: As soon as valuable new materials were created in space, research laboratories reacted to the challenge to replicate them on earth at a fraction of the cost.

One of the most spectacular changes in materials came from the continuing improvement of metallic glasses. Molten metal alloys are sprayed at high velocity onto a chilled metal wheel that rotates rapidly. When the molten alloy touches the rim, it "spot-cools" into a solid ribbon in less than a thousandth of a second. This leaves the molecules arranged in random patterns rather than in the form of interleaved crystals, as in conventional alloys. The metallic glasses are extruded at many thousands of feet per minute.

VIDEO: The violent hiss of metallic glass extrusion machinery.

NARRATOR: These are toughest, most corrosion-resistant materials known, lighter than aluminum and far stronger than steel. They are woven into fabrics, braided into tubes, and laminated into plates.

VIDEO: Scenes of third-world starvation. Children with pot bellies and skin stretched over fleshless bones.

NARRATOR: The fear in the 1980s was that no matter how much technical ingenuity the first world showed to get around their problems of power and raw material shortages, the huge problem of the third world's starving millions remained. Conventional agriculture was simply not enough to feed the coming world population of 8 billion. And to make matters worse, land useful for agriculture was being lost at a rate of 27,000 square miles per year. Additional millions of square miles were threatened by the spread of deserts. One school of thought advocated a back-to-nature movement, with everyone growing food in the backyard. But the land available for this was not nearly enough. It could not have fed one billion, let alone 8 billion. The population is too large.

Throughout the twentieth century, technology became more and more important to agriculture: machines to do the work, chemicals to affect the soil and the environment.

VIDEO: We see the narrator walking through a hydroponic farm. The plants are enclosed in translucent plastic bubble-like greenhouses. The plants grow without soil in plastic troughs 30 centimeters wide, down which nutrients trickle. The plants are tied to strings for support. Most of them need no overhead watering. Their roots are exposed in the liquid of the troughs. At the end of each trough, 25 meters long, the nutrients are analyzed electronically, the moisture adjusted, and the liquid pumped back to the head of the trough.

NARRATOR: By the 1970s it was in farming that it was most absurd to say, "No more technology." Without technology it would have been impossible to feed the current population of the world, let alone the huge increase of population anticipated by 2020. The bioengineering and hydroponics research was vital.

> *VIDEO*: The narrator picks tomatoes and eats one.

NARRATOR: They taste great. They don't taste as though they were grown with unnatural fertilizers.

> *VIDEO*: We see a rich crop of tomatoes, cucumbers, lettuces, and other vegetables.
>
> More pictures of famine in India, the Ogaden, and African deserts.

NARRATOR: The fears of world starvation, vividly expressed in Victorian times by T. R. Malthus, reemerged.

But this was a problem solvable by technology. It was tackled with a growing sense of altruism and urgency by universities and government research centers. Before long, industry began to perceive the new aids to food production as an area of high technology. It was attacked from all directions.

> *VIDEO*: The camera tracks along lengthy hydroponic farms with vegetables growing in racks fed by chemical nutrients. Lettuces grow on vertical styrofoam production boards that lean together. Tomato plants with large ripe crops hang from the ceiling. Watercress grown on styrofoam floats in a pool where catfish feed on its roots and fertilize the nutrient-high water. Melons are ripening on A-frame trellises above the water. Bush beans, pole beans, and corn are planted in alternating rows. Large inflated polyethylene greenhouses house this rich collection of plants, in the desert and by the seashore.
>
> The density of plants is high, and the fruit lush. We look down mile-long avenues of hydroponic crops filmed into the sunlight. Troughs of different vegetables are stacked one above the other.

NARRATOR: Many plants can be grown more efficiently if they are not grown in soil. Carefully controlled nutrients are fed directly to their roots. This hydroponic lettuce farm grows 2 million heads of lettuce per year per acre. If the same lettuces were grown in conventional soil beds, the yield would be only 20,000 per year.

> *VIDEO*: The camera pans along the troughs. We see the exposed roots of the plants growing in shallow trickling liquid. Robots tend the crops, with gardeners supervising.

The feeding of the plants is accurately controlled by microcomputers to give the highest yield. The glimmering sunlit images give a sense of lush fruitfulness.

NARRATOR: Much research went into selecting and crossbreeding the most efficient varieties of plant. Genetic engineering made it possible to insert the protein-forming genes of one plant into another. Gene transfer techniques made possible much higher food yields.

Photosynthesis—the process by which plants convert the energy in sunlight into plant sugars—was greatly improved. Traditional plants convert only about one percent of the energy absorbed. Many of today's plants convert as much as 4 percent, which quadruples the rate of growth.

NARRATOR: (over video) A third of the earth's surface is desert; even more is uncultivated scrub land. These areas are mostly ideal for hydroponic farming because there is plenty of sunlight. Solar energy powers the electronic equipment used to control the nutrient feeding. The low amount of rain that falls is stored where it cannot evaporate, and plants are grown that need no aerial watering. Alcohol-producing plants are grown, and fuel made locally powers the heavier machinery.

VIDEO: Helicopter shots panning across coastal desert.

NARRATOR: Twenty-two thousand miles of the earth's coastlines used to be desert. A particularly important farming development was halophyte crops. These are plants that can be irrigated with seawater. Such plants are used to help reclaim desert land near the sea and provide vast quantities of livestock forage and food for people.

VIDEO: Pictures of a vast halophyte farm on the coast of Mexico.

NARRATOR: The growing of food has become a high-technology operation. Much of the technology came from the United States. In 1970 the U.S.A. exported 6 million tons of food and was regarded as the breadbasket of the world. Today it exports 30 million tons.

As genetic engineering and hydroponic techniques spread, high-technology corporations moved into agriculture. High-tech companies built vegetable factories and hydronics control equipment. Hundreds of seed companies were bought out by the chemical and drug companies like Monsanto, Dow, and Eli Lilly. These companies spent vast sums on genetic research and gene splicing. New genetic products began to replace many of the cruder herbicides, pesticides, and fertilizers.

To make protein, plants need nitrogen. Bacteria that grow on the roots of plants take nitrogen and combine it with oxygen and other substances to produce nitrates, which nourish plants. For decades crude artificial fertilizers were used, but these eventually weakened the natural bacteria. That meant more fertilizers and more expense. By using genetic engineering, it became possible to infect plants with superior nitrogen-fixing bacteria, which nourished the plants, improved their yield, and saved billions of dollars in world fertilizer bills. This became a vast, profitable industry.

Plant regulants—chemical-like hormones that alter a plant's growth and quality—were created. These helped scientists to modify crops selectively. For example, the sugar content of cane sugar was increased enormously. Nut trees were made to bear four times as many nuts.

Oxygen inhibits growth in many plants, so some plants were grown in robot-controlled greenhouses flooded with carbon dioxide.

VIDEO: Plant-tending robots at work on a row of plants 1000 feet long. There are several levels of plants, each level with its own robot track.

NARRATOR: Shipping food from sunny climates vast distances to the consumer made sense in the 1960s when gasoline cost 25 cents per gallon. Now transport is expensive and food is often grown hydroponically at the consumer's location.

Plants are grown to create fuel. Euphorbia plants are processed for the oil in their stems and leaves. Special breeds of sugarcane are grown in great quantity to produce alcohol fuel. Water hyacinths grow prolifically and create methane gas. In many parts of the world, land regarded as nonfertile 50 years ago is now growing food. Desert reclamation has progressed a long way since the Israelis first achieved spectacular results in the 1960s and 1970s.

VIDEO: Helicopter shots racing across infertile scrub land in Africa, dissolving to desert reclamation projects.

NARRATOR: In the bright sunlight of the desert, various means were found to create cheap power. With the power it became possible to desalinate the ocean and pump the desalinated water hundreds of miles along pipelines to irrigate vast desert areas.

VIDEO: We see the work of reclaiming the Australian desert, much of it similar to the Israeli work, only on a far grander scale, much of the work being supervised by Israeli scientists, many exiles from the Middle East.

NARRATOR: Not only were orchards created in what had been desert, as was done in Israel, but huge hydroponic plant factories under domes were also built. Nearly all the work in the domes is done by robots. What was once totally unproductive land has now become a huge food supplier to the world.

The presence or absence of trees has a major effect on potentially arid climates. In Africa the wholesale cutting of trees for firewood was a major contributor to famines that devastated parts of the continent. Elephants and goats added to the destruction, leaving large areas devoid of trees. Trees produce oxygen, which helps to cause rainfall. Their leaves fall and build up humus in which other plants can grow. Where there are no trees, high winds sweep across the plains, and the wind and blazing sun parch all but primitive vegetation.

Ecomanagement, where it existed, harnessed the wind. Cheap, reliable, two-blade windmills pumped water from deep underground. Fast-growing, resilient trees were planted first, then more robust, long-lived trees, which were better for humus buildup.

VIDEO: A sequence of scenes flash on the screen, labeled "Year 0," "Year 1," "Year 2," and so on.

"Year 0" shows rolling desert in Northwestern Australia with scrub desert vegetation. The wind howling on the sound track blows sand across the desolate landscape.

"Year 1" shows windmills, protected water channels, and plantations of small casuarina trees.

"Year 2" shows grass near the water channels, sprinklers at work, patches of vegetables.

"Year 5" shows the area covered in grass, wildflowers in the grass, cows grazing; the casuarina trees are 16 feet high, and other, more slowly growing broad-leaved trees have been planted. Some humus is beginning to build up.

"Year 10" shows many trees. The casuarinas have trunks a foot thick and are 30 feet high. Houses and domed greenhouses. Robots tending crops.

"Year 15" shows lush orchards, complex and diverse vegetation, rich farmland.

Next we see some of the technology that made the ecology of northern Australia less savage to humans. We see the radar system used to scan the sea for sharks, the various means of controlling harmful insects by breeding predators, and wind farms with 100 wind generators creating electricity to pump water.

The sequence builds to a climax of lyrical beauty: the desert blooming with carpets of flowers of all colors, fruit orchards and garden cities where once there was nothing but sand and grit.

Cut to the narrator in her roof-top garden, which contains not only flowers but also fruit trees and vegetables in tropical abundance.

NARRATOR: One of the fundamental beliefs of our society is that we must husband and take care of the planet. It is the only decent one we have. We must not create dust bowls and deserts. The nonfertile areas must be developed slowly with the right mix of trees and plants until these areas are fruitful. This attitude of mind is a relatively recent one, one that we were forced to adopt after the crises of earlier decades.

VIDEO: We see a huge used-car graveyard with the large American cars of the 1950s and 1960s, with chrome and tailfins. Cut to other dump sites and landfills of the 1980s and 1990s.

NARRATOR: It is difficult for us to believe how much waste there was in the processes of the industrial nations in the 1970s and 80s. An average middle-class home generated enormous quantities of garbage: paper and plastic packaging; clothes discarded because stitching had come undone, though the rest of the garment was perfectly serviceable; tools and gadgets thrown away because they would cost too much to mend.

Industry thrived on planned obsolescence. It planned goods carefully so that they would have to be replaced in a few years. Machinery had plastic gears or catches designed to wear out after an appropriate time; light bulbs were designed to have to be replaced after 1000 hours or so; synthetic fabrics with spun fibers (unlike normal nylon) had good properties but were not used because they lasted too long. Clothes were made obsolete quickly by carefully manipulated changes in fashion. If people kept goods too long, profits would be lower.

VIDEO: This sequence builds to a climax of throwaway waste: discarded clothes and furniture in the streets of New York; the deliberate destruction of food surpluses to keep prices high; thrown-away washing machines and refrigerators.

NARRATOR: The life style of excessive consumption of manufactured goods spread beyond the major industrial countries as television slowly penetrated the third world.

VIDEO: A poor Indian village with mud and lath houses, cows in the street, villagers sitting in doorways. The camera closes in on a TV set in the village square. It is showing the program *Dynasty*.

NARRATOR: Third-world countries cannot afford to make much television of their own. They buy mainly American television programs. For some years the show *Dynasty* was the most popular.

VIDEO: We see the extreme contrast between the poverty of the village and the life style on the screen.

NARRATOR: Third-world material expectations increased rapidly as communications satellites and earth stations brought television everywhere, so that it became in the interest of multinational companies to advertise, advertise, advertise to their third-world audience and potential market.

VIDEO: A television broadcasting satellite. Advertisements for Japanese products being shown in the Indian village.

NARRATOR: The problem with much of the third world's food production is political, not technical. We have helped much of South America to solve its food problems, but other parts of the world are still in chaos.

VIDEO: We see appalling scenes of starvation in an African semi-desert region. This is intercut with lyrical sequences of the Australian desert blooming and hydroponic greenhouses with rich crops.

NARRATOR: The technological problems of almost unlimited food production have been solved. Food production has caught up with population growth. But there are still millions of starving people in India and Africa. The problem is not technological but sociopsychological and political.

VIDEO: Aerial shots of India from a plane flying low and fast.

NARRATOR: As you fly over India or Africa and look down, still today most of the land is not being used to grow food efficiently. The efficiency of farming depends to a great degree on the amount of power available. With an increase in power, there can be an increase in food production. India and Africa have abundant sunlight, and sunlight can be used as a power source. With power, water can be brought to the desert, and once there is water, vegetation. Water and vegetation rapidly bring a change in the ecology.

VIDEO: Helicopter shots racing across the northern Sahara, still endless rolling sand dunes.

NARRATOR: Much of the world's desert was made by humans. What is now the northern edge of the Sahara was the Roman Empire's equivalent of today's Canadian prairies—the granary of the world.

VIDEO: Ancient pictures of Roman grain in the Sahara; corn in Egypt.

Large fields in Kenya with many women hoeing the soil with manual hoes, some of the women with babies in slings on their backs.

NARRATOR: Much of Africa is still more primitive than Europe 1000 years ago. People there do not even use beasts and plows to break up the soil. Near to this manually-hoed field is a Ministry of Agriculture building with 200 people working in it, but there is little attempt to introduce sensible technology.

VIDEO: A studio discussion among the narrator, an Indian doctor, and an American.

American: It's entirely a management problem. If India works at its problems in its own way, millions will continue to starve. If we offer technical advice, it will be used wrongly or ignored. If we give food away, it will increase their dependency on us. If we sell it, they can't afford to pay for it. The only satisfactory way to feed India would be to conquer it and run its food production ourselves. This we have said we will never do. Are we wrong?

Indian Doctor: Far be it from me to recommend imperialism again.

Narrator: Which would you prefer: your country ruled by advanced nations and your people fed? Or your country starving?

Indian Doctor: India can never be like America and certainly does not want to be. The world will benefit greatly from having diverse cultures. But we need some better means to transfer management skills. Most world starvation is a management problem today. But Indian problems must be tackled within the context of Indian cultures.

A basic principle of evolution is that extinction results from lack of diversity. If the whole world was homogenized by your wired society, that might be disastrous. It's possible that the Hindu way of life will survive long after your way of life has destroyed itself. It has survived all the oppressive regimes that have ruled India.

NARRATOR: The computer models warned that the world was heading for disaster: World population was growing exponentially; the use of resources was growing exponentially; the earth's resources were limited; the earth's resources were running out.

After recurrent energy crises and worsening shortages of raw materials, the warnings were heeded, and a different vision of society began to develop. Materials can be recycled, waste can be put to use, goods can be built to last longer, energy can be saved, and renewable sources of power such as sunlight can be harnessed. A throwaway society that consumes ever increasing quantities of nonrenewable resources is being converted to a recycling society that conserves its resources where possible.

VIDEO: Walt Disney World. Mickey Mouse figures among the crowds. Children screaming on rides.

NARRATOR: By 1980 Disney World was the first large community to have developed a closed system in which all waste was recycled.

VIDEO: We go underground at Disney World, following the garbage as it is sucked down, then moved to a furnace, which generates hot air. We see details of the vast concrete tunnels, the computer rooms, the wardrobes, laundries, and administrative offices underground, all hidden from the visitors above.

NARRATOR: Many industrial processes create heat. In the 1970s this was mainly all wasted and frequently caused damage to the environment. Waste heat can be used to heat homes or greenhouses.

VIDEO: A power station with large cooling towers and many hydroponic greenhouses like those shown earlier.

NARRATOR: In 1970 the waste heat from power stations was thrown away. By 2000 it was put to good use.

VIDEO: 1970. Sewage being pumped into the sea, fouling the water.

NARRATOR: To waste sewage was to waste a most valuable asset.

VIDEO: Various devices in use in third-world countries to turn human and animal excrement into methane gas and fertilizer.

NARRATOR: Increasingly, industry had to operate in a closed system, recycling its waste products and using or selling its own waste heat.

VIDEO: An industrial complex of 2019 near the sea. Most of the factory facilities are underground. Aboveground are gardens and greenhouses. The greenhouses are higher than normal, with raised walkways giving access to the higher-up plants. There are different types of plants at different levels.

We tour the complex as the following discussion progresses.

NARRATOR: The heat from the underground factories is piped into the greenhouses. The plants in these greenhouses are highly interdependent. It is a polyculture system. Beans produce their own fertilizer from nitrogen in the air, but corn can't do that. The corn is therefore fertilized by the beans. In turn, the cornstalks act as poles for the beans. Intercropping allows different crops to help each other.

The saltwater channels are used for shrimp farming. The salt water evaporates and condenses on the polyethylene, where it flows into the tanks ready for nutrient mixing. Trailing roots and leaves help to feed the fish. We harvest 85,000 pounds of shrimp from each acre of water. That is more than a season's catch for a fleet of ten shrimp boats, and it is grown at a fraction of the cost. Where freshwater fish are grown, these in turn provide nutrients for the plants.

Each greenhouse has a complete self-contained ecology of its own. The vertical growing system causes us to measure the yield per *cubic* foot, not the yield per *square* foot as with conventional agriculture.

The factory complex and workers' homes are in the same location to eliminate unnecessary travel. Because the factories are underground, the environment can be made very attractive. Heat from the factories goes to the homes. Sewage from the homes goes to the greenhouses. The homes have garbage chutes. The garbage is sifted by machines, and the combustibles are used in the factory furnaces. Metals are recycled. The ash is used as fertilizer. A separate garbage chute is used for food products, and these go to computer-controlled compost units. The compost goes to the greenhouses, and householders can use it for their own gardens. The roofs of the houses catch rainwater, which is piped into the drinking-water system. Water from washing machines and bathrooms is reused by the factories and lawn sprinklers.

NARRATOR: In 1970 solar electric cells were small and very expensive. They were attached laboriously to the surface of the early communications satellites. Only in space was it economical to use this form of electricity generation.

> *VIDEO*: Men in white laboratory coats building the outer shells of the Hughes 333 satellite. Dissolve to a picture of that satellite in orbit, accompanied by a babble of telephone sounds.

NARRATOR: But solar electric cells dropped in cost as fast as the cost of silicon circuits for computers. Amorphous silicon panels were built, rapidly improved, and later mass-produced. The roofs of buildings were redesigned both to collect water and to hold solar-electric panels. Solar panels on vehicles recharged their batteries.

VIDEO: Solar panels on roofs of houses, Citibank in New York and on sloping skyscraper roofs in Chicago. Factories with high solar arrays designed to rotate with the sun. Solar panels on cars.

NARRATOR: Legislation was necessary to control both pollution and the waste of resources.

During the 1970s and 1980s auto emissions were brought under control, and lead in gasoline was banned.

The laws governing power authorities were changed to permit complexes in which the waste heat and ash were put to good use.

Windmills of new design were very efficient. The electricity grid was adapted so that homes or town windmills and solar systems could feed the grid when not using all the power they generated. The grid became a distribution grid rather than merely a broadcast grid. Laws mandated that power companies must buy electricity from householders and town facilities at a price that was their own minimum cost of production.

VIDEO: Home wind generators. A California wind farm with hundreds of large-blade generators on a forest of tall poles.

NARRATOR: The trading of energy became a much more complex process, but the complexity of detail was easily handled by computers, which now allowed for many different producer-consumer relationships.

Planned obsolescence was legislated against by requiring five-year guarantees on all products unless there was good reason otherwise. Later the law demanded ten-year and then 15-year guarantees.

As we increasingly learned to control our environment, the pressures of population grew. It took most of human history to learn how to feed a billion people, and then the earth's population grew to 2 billion, 4 billion, 8 billion. The demands for affluence, energy, large homes, travel, and factories strained the planet's resources more than the struggle to grow enough food. Wilderness disappeared and was replaced by the artificial: spreading suburbs, overgrown cities, mechanized farms, roads, recreation areas. Nature was replaced by the creations of man.

For hundreds of millions of years the forces of evolution shaped the planet. Plants and animals evolved, adjusting themselves over millennia. Nature adapted slowly with seemingly infinite patience. What will happen to evolution now?

NARRATOR: (over film footage) As human creations spread, the wild animals had to be controlled. In Africa one could not have lions in the villages or giraffes in the orchards. The hippos and antelopes had to be prevented from eating the crops. An African elephant does 3 tons of damage to trees each day. Large ditches and fences were dug to confine wild animals to game parks.

In the twentieth century wild animals were destroyed at a furious rate. First there were big-game hunters. One British government employee in Africa boasted of shooting 971 rhinoceroses. The big-game hunters became mechanized. Few things were easier than to shoot a lion from a four-wheel-drive vehicle. The animals do not attack vehicles and usually do not run away from them, not instinctively recognizing them as enemies or even as creatures. In a way it was easier to shoot lions or rhinos than to shoot rabbits.

> *VIDEO:* Western magazines showing models in leopard coats. A Chinese drugstore showing on the shelves items for increasing sexual potency, such as powdered rhinoceros horn and tiger's penises.

NARRATOR: When shooting for sport was stopped in some countries, killing by poachers continued. Rich people in the West paid large amounts of money for fur coats; rich people in the East paid even larger sums for the horns of rhinos and certain antelopes in the belief that they increased male sexual capability.

NARRATOR: (over video) The armies in Africa's endless wars killed the animals in the game parks for food. As the population grew, increasing numbers of animals were killed for food, and the game parks were restricted to limited areas.

Tourism became the largest source of foreign revenue in some African countries. As air transport became cheaper, tourists flooded to the game parks. Busloads of tourists scoured the parks on game drives every morning and evening. The tour operators wanted the animals to be easier to find. They built water holes filled by wind-operated pumps, and in some places built hotels on pillars overlooking the water holes so that tourists could watch the wild animals coming and going while cocktails were served. Underground observation rooms were built for viewing hippos from underwater.

With the second generation of supersonic jets, rich people went on weekend "safaris" to Africa and elsewhere. Colorful tourist balloons drifted across the game parks. The elephants ignored the balloons, but lions and other animals would often run from them and hide. The balloon operators would attempt to bring their balloon down in herds of giraffe or zebra. Later, because the balloons carried few people and were at the mercy of the wind, game park dirigibles were built, and far more tourists could watch the wild creatures from above with telescopes provided.

Some game parks sold the adventure of dusty drives in open land rovers. Others operated monorails with rubber-wheeled vehicles gliding through the bush to carefully planned water holes. Elevated monorail tracks in thick jungle wound through the trees, enabling tourists to watch for monkeys and gorillas. Tourists in air-conditioned comfort clicked their computerized cameras.

VIDEO: Tourist advertisements for luxury trips to Africa. "See the animals in air-conditioned luxury."

NARRATOR: Africa lost its monopoly on game parks. Relatively small artificial game parks were operated in Europe and Japan. Australia and eastern Asia set up vast parks with African and Asian animals, to enhance their tourist trade. Some islands were converted to game parks so that visitors could explore among the animals without being endangered by lions or other potential killers.

The future evolution of species has become a subject for debate. It is clear that we are interfering with natural evolution in a massive way.

It is difficult to imagine how evolution, if at all, will respond to the busloads of tourists. A few animals born in the parks seem to sense that they are supposed to perform for the visitors. Polar bears in some locations put on extraordinary performances for the tourists at underwater viewing windows.

Nature's evolution of the genus *Homo*, on the other hand, is being massively interfered with with all manner of synthetic drugs—psychotropic drugs, antipsychotropic drugs, contraceptive pills for both sexes, morning-after pills, memory-enhancement pills, drugs for high blood pressure, and so on. Many drugs in widespread use affect emotion and sexuality. When we experiment with emotion and sexuality, we experiment with our own species' future.

Humans are creatures in the process of evolutionary change. There is evidence that the species is in a transition stage groping toward better adaptations at a faster rate than most creatures. Yet our use of drugs drastically changes courtship patterns and emotional needs. To control population growth in the third world, widespread use has been advocated of pills that combine contraception with enhanced sexual pleasure. Drug manufacturers are experimenting on a massive scale with drugs that enhance intelligence and that slow aging. There are drugs for controlling aggression, drugs for increasing assertiveness, and massive use of artificial vitamins. The money spent on mood-altering drugs far exceeds that spent on computers. We are tampering with biohistory on a grave and grand scale.

> *VIDEO*: Montage of microscope gene splicing shots and television drug advertising.

NARRATOR: Meanwhile, the evolution of our mechanical and intellectual capabilities is changing at a rate millions of times greater than natural evolution. The evolution of intelligent machines, robotics, and worldwide computer networks seems destined to change ever faster as computers become more powerful, more highly parallel, more "intelligent," and increasingly self-programming. The proposals for advanced machinery in space suggest artificial habitats competing with the earth itself. Our capability to destroy nature is growing frighteningly. It is as though all nature has become our backyard, which we can cultivate or neglect at will.

Meanwhile, we do not know whether we will avoid nuclear war. In the long run it seems possible that the stockpiles of weapons may be used. It seems possible that a two-year nuclear winter may occur. The entire planet may be devoid of life like a garden of house plants that has been left unattended.

> *VIDEO*: Dead house plants. Dissolve to nuclear winter devastation. Cut to scenes illustrating the following discussion.

NARRATOR: So massive preparations are being made. Seeds of as many plant species as possible are stored, indexed, and regularly replenished. In the southern continents there are several Noah's Ark projects where animals would survive until sunlight and vegetation return. In the fjord area of New Zealand electric stations have been built deep in the mountain rock, powered by water tunneled from alpine lakes 1000 feet above. In the underground chasms are supplies of food and goods to support hundreds of people for several years. Washington, Moscow, Canberra, and Pretoria all have their own nuclear-winter-proof shelters.

NARRATOR: As we speculate about other civilizations that we believe exist in vast quantities far away in the universe, totally out of our reach, themselves isolated by unimaginably vast distances, it seems likely that there also a species would have eventually acquired the intelligence to achieve mastery over the natural. They too will have built machines and networks and satellites. They too may have reduced nature to a backyard cultivable or destructible by technology. They too will have had wars and accidents and a turbulent passage to maturity.

Perhaps evolution was designed so that this should happen. The inventions of nature grow and compete. Survival of the fittest dominates evolution until one species learns science and engineering. It quickly dominates its planet, perhaps destroys its planet, but clings to life. It builds and experiments and learns to control both its habitat and itself. Evolution changes from the dismally slow adaptions of nature to something millions of times faster. An intelligent species invents, builds, and unlocks the secrets of molecules. It creates intelligent machines that intercommunicate at the speed of light around the planet. The machines beget other machines. There is a chain reaction of intellectual invention.

There seem to be two thresholds following fast one upon the other. First a species learns to invent the artificial. It creates engineering—mechanical, chemical, biological, nuclear, and genetic. It has its industrial revolution. It amplifies its muscle power a millionfold. It can play havoc with its planet. It learns by trial and error, like a child, to master its new power. Some of its errors are painful, but it recovers. The crudity of evolution in the jungle with creatures eating other creatures is replaced by evolution in laboratories, board rooms, and law courts.

Second, the machines become intelligent. They can help create new machines at great speed. They intercommunicate. The first threshold amplified muscle power; the second threshold amplifies intellectual power. The computers can reason, calculate, and create designs billions of times faster than their biological creators. But the biological creator remains the master, and the machine, however powerful, remains a tool.

It seems as though nature were designed so that this evolution of artificial intellect should occur. It occurs completely differently on the different islands of life that develop in the galaxy. These islands are isolated from one another by uncrossable distances. The isolation, as though deliberately planned, causes a great diversity of different forms to evolve on different planetary systems. Because they cannot communicate, no one island can dominate the others. Each is an isolated experiment in evolution.

VIDEO: A black void with stars. The camera moves slowly through the stars. Individual stars flash and pulsate, suggesting frenzied activity.

NARRATOR: Some of the experiments fail, just as some of evolution's experiments on earth have failed. The isolation prevents a failure on one system from bringing down other systems.

In primitive evolution, survival of the fittest is a powerful but brutal means of advancing the species. It ceases to function when one species becomes dominant and makes itself comfortable. A different form of evolution takes over—a species wrestling with the products of fast artificial development. If a nuclear winter occurs, only the stronger members of the species will survive to rebuild. We may still have survival of the fittest.

In any case, the planetary system is now in our care. It seems totally and indefinitely isolated. We can travel around our solar system but no further. Our technology is evolving furiously. Nature probably has further means of ensuring advancement of the species through survival of the strong.

VIDEO: Zoom out from the planet Earth in a black void.

NARRATOR: We are on our own.

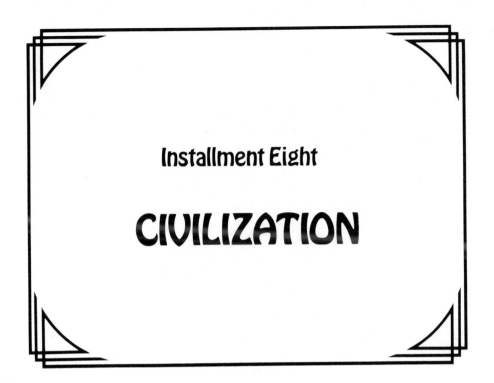

Installment Eight

CIVILIZATION

PRESENT-DAY AUTHOR: In the factory of the future, robot production lines will create vast quantities of complex products. Automation, genetic engineering, and other advancing technologies will make society steadily wealthier, and humans can, if they choose, have more leisure time. How should people live when society is rich and jobs are automated? What should be the character of a society much wealthier than that of today?

Often a person who works hard and becomes very wealthy does not know how to spend the money in ways that are really worthwhile. Such people are much better at making money than at spending money. The same is likely to be true with societies as they evolve from struggling economies into mass industrial societies and then into postindustrial affluence. How should Japan, for example, spend its rapidly growing wealth? Today it is impressive for its mastery of automation but not for its quality of life.

There is no question that the technology we are unlocking is going to change society beyond recognition. But can we change it for the better? And if not, what is the point of technology?

Given the awesome technological opportunities, everyone ought to be reflecting on the question of how we should put technology to use. What should be the characteristics of a great civilization?

This is perhaps one of the most controversial questions for a philosophy of technology, but it is surely one of the most important. A sense of values is as important to society as genes are to a complex creature. There is likely to be substantial agreement about the lower building blocks of that sense of values, but the highest block relates to the difficult question, What does it mean to be fully civilized?

Few people seem to be asking that question today. Ours is an age largely without philosophers. Some universities have expressed worries that they are turning out technological barbarians, but they do not seem to take any action about this worry.

We have stressed the importance of having a vision of alternate futures. Only if we have this vision can we attempt to steer a course. Inventing the future is much too important to be left to engineers and marketing people. It needs to be a debate involving intelligent people of all callings. The vision needs to be firmly in the minds of the public; it needs to be on television. Today television distorts the vision of our future appallingly.

Goals in institutions do tend to be met if everyone understands them. We need to know what outcomes we would like from our breathtaking race into future technology. A distorted vision of the future is dangerous.

We have emphasized that a basic rule should be that human potential is improved at the same rate as technological potential. In the short term this human potential relates to people doing different work. The rapidly changing job mix requires more use of intelligence, more creativity, more skills, more self-sufficiency. The economy requires more entrepreneurs. In the longer term the improved human potential relates to civilization itself. If we are to build a golden age, what human attitudes and understanding will it need? If the future is to be an age of greater leisure because we have automated our offices and factories, we should be educating children to use that leisure well—to create music, to appreciate art and literature, to have enjoyable physical skills like riding horses and skiing, to create theater, to be literate, articulate, witty, and informed.

The most important question of today is a question that every intelligent person ought to be debating: Given the ability to work miracles with technology, what sort of world do you want your children to live in?

VIDEO: *The same scene-changing sequence as before is used to slide from the present to A.D. 2019: a rushing* *sensation like flying very low at extreme speed through a mountainous valley, but the earth and hills look like endless etched microelectronic circuitry. Electronic music is used, evocative of time travel.*

NARRATOR: Project Jacob is a scheme for dramatically lowering the cost of transportation into space. It would undoubtedly open up new frontiers in space. However, it is outrageously expensive if we compare it with what the same amount of money would buy if we spent it on enhancing our ability to be civilized—if we spent it on the arts, drama, worldwide videodisk libraries, or Michelin three-star restaurants.

> *VIDEO:* The camera tracks back and we see that Corinthia is in a superbly elegant three-star restaurant in Paris. The waiter is taking an order with Parisian finesse.

NARRATOR: It seems less expensive when we compare it to the costs of defense.

A furious worldwide debate has been asking the question, What is the point? Surely the money can be spent in better ways. The same argument could have been applied to the 1960s, when it cost the United States almost one percent of its gross national product (at peak) to put the first humans on the moon.

> *NEWSREEL: Saturn V* blastoff at Cape Kennedy in 1969. The first step onto the moon.

NARRATOR: Our history books record Project Apollo as a great step for humankind; they do not ask the question, Could the money have been spent better? The avid proponents of Project Jacob say that history books will record this also as a giant step. More money spent on culture, they say, could be done at any time, but Project Jacob is our manifest destiny *now*.

The debate relates to civilization. What does it mean to be civilized? We are a *rich* society. How does one spend that money to be a great *society*, to become one of the golden ages in the history of civilization? What do we mean by civilization's golden ages?

VIDEO: The waiter walks to the kitchen. The camera follows him through the kitchen door, and we see that the food is cooked almost entirely by robots.

NARRATOR: The discussion of whether $200 billion should be spent on Project Jacob has tended to add fury to the pro-technology and anti-technology arguments.

Project Jacob is a long-range scheme for achieving transportation into space by means of cables to orbital locations. Vehicles climb the cables into orbit. Project Jacob is named after the story in Genesis of Jacob's ladder.

VIDEO: A preacher reading the Bible at a Medieval lectern.

Preacher: And Jacob dreamed, and behold, a ladder set up on the earth, and the top of it reached to heaven: and behold, the angels of God ascending and descending upon it. [1]

NARRATOR: The project consists of two stages that are fully designed. Two additional stages are foreseen for the future but are beyond the capabilities of today's materials.

The first stage is referred to as a space catapult.

ANIMATION: The catapult is demonstrated as it is described.

NARRATOR: A continuous cable spans two Pacific islands near the equator, 1600 kilometers apart. The cable, originally supported by large balloons, is rotated increasingly fast until it reaches a speed at which it is pulled away from the earth by centrifugal force.

When it reaches its final position, its ends rotate through coils acting like pulleys on the two islands and its middle reaches a height of 385 kilometers (see Fig. 8.1). The cable passes through an evacuated tube that is held motionless with respect to the ground. It is kept spinning by linear induction motors. The "pulleys" on the islands are in fact rings of superconducting magnetic coils so that the very high speed cable never makes physical contact with other components. The very light braided aluminum alloy cable has induced currents of high power because of its speed. The currents interact with the

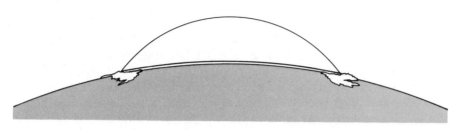

Figure 8.1

magnetic coils. A nuclear plant at each island provides the 200 mega-watts needed to power the cable.

Payloads are carried along the cable by means of sleds consisting of superconducting coils mounted on a titanium frame. These can be linked together into long trains for carrying heavy loads. The sleds accelerate up the spinning cable, reach its maximum velocity, fling their contents into orbit, and then ride down the cable to the opposite ground station. They can return via the other loop of the cable. [2]

NARRATOR: The second stage is called Jacob's ladder. Two rings of substantial mass rotate around the earth in opposite directions in a low orbit. The space catapult is used to take the materials for the second-stage rings into orbit. [3]

ANIMATION: Jacob's ladder is demonstrated as it is described.

NARRATOR: Suspended from the massive rings down to earth is a cable. At the top the cable is attached to a supporting structure that rides along the rings and is held up by superconducting magnetic coils. Linear induction motor drive coils help the structure to ride along the rings.

The two rings rotate in opposite directions, and the cable support is attached to both. The drag forces in opposite directions cancel each other. In case of failure, numerous independent coils are used for each cable support. Eventually, multiple cables would go to each space station complex.

The cable must be strong enough to hold both the weight of the cable and the vehicles that travel up and down the cable. The cable is almost 300 kilometers in length, and a graphite whisker mate-rial of great tensile strength is needed for the cable itself.

The stations at the top of Jacob's ladder would be still relative to the earth and would be 300 kilometers high (as opposed to geo-synchronous satellites, which are 35,000 kilometers high). These sta-tions would be used for broadcasting television and relaying telecommunications (in the daylight hours). A signal from them would be 10,000 times stronger than the same signal from a telecommunications satellite of the same power. A signal that they receive from earth would also be 10,000 times stronger. This would dramatically lower the cost of broadcasting very high resolution TV and other signals.

A third and fourth stage are visualized for the future. They could not be built today because we do not have materials strong enough. We make fibers today with tensile strength ten times greater than 30 years ago. It is expected that future fibers will have a further improvement in strength of a factor of 10. Then cables will be linked to stationary objects in geosynchronous orbit rather than to moving rings. Such a cable would have to be extremely strong in order to support its own weight. The cable would be tapered so that it is thickest at the top, where the load is greatest.

ANIMATION: The third and fourth stages are demonstrated as they are described.

NARRATOR: The third stage is to link a cable from the geostationary station riding on the orbital rings in low earth orbit up to geosynchronous orbit.

The fourth stage is to build a cable directly from earth to geosynchronous orbit. This would be simpler and more elegant. It would bypass the need for massive orbiting rings. However, it could not be done until much stronger and lighter cable materials are produced. Project Jacob calls for intensive research in this area.

To launch the third and fourth stages a very fine thread would be linked from the earth vicinity to geosynchronous orbit, and the larger cable would be lowered a stage at a time down the thread.

It is estimated that the cost of taking people and goods to low earth orbit with stage 2 would be less than a thousandth of the cost with the space shuttle of the 1980s. The construction costs would be very high. The return would come from cheap power, mass production of molecular electronics, biochips, exotic alloys and pharmaceuticals, cheap telecommunications, and military deployment of beam weapons.

The orbital rings, if built, would be a resource for the whole earth, and it is expected that many countries would operate their own stations on the rings. The project has been organized for multinational funding and management, and the controversy about the project has been raging on television channels around the world.

We are concerned here with civilization, its meaning, and how it can be improved by technology. Can the goals of being civilized be achieved better with the aid of such a project?

VIDEO: Various spokespersons in the debate that follows start their comments in a window created electronically behind the narrator. As they talk, the window expands to fill the whole screen.

NARRATOR: (to camera) The debate about Project Jacob is a reflection of a much larger concern: How should a wealthy society spend its money? Given the wealth and leisure created by the machines that surround us, we ought to be creating a golden age in which humanity aspires to be truly civilized. But what does this mean?

MUSEUM CURATOR: (in a screen window behind the narrator) A society with a highly developed sense of excellence in all things. A society that honors excellence.

UNIVERSITY CHANCELLOR: A society that gives endless opportunities to those that can take advantage of them.

PSYCHOLOGIST: But also a compassionate society. A society that makes a deliberate legal place for compassion.

HEAD OF PROJECT JACOB: A society that challenges humankind. A society with challenges on a grand scale that stretch the frontiers of humanity into hitherto unknown areas.

STUDENT: A society filled with excitement where boredom is crime.

NARRATOR: Civilization implies the search for pleasures that are subtler, removed from and more refined than those to which brute instinct leads. A savage makes one of his first steps on the way to civilization when instead of eating a rabbit raw, he takes it home and cooks it. He has a long way to go before he creates great restaurants.

> *VIDEO*: We see the three-star restaurant, a robot cutter dicing shrimps on a cooking grill at the side of the table while a nozzle sprays them with sauce. The machinery gives the impression of great precision in the food preparation.

NARRATOR: It is a long way from the native's song and dance to the appreciation of Mozart and Stobart's production of *Parsifal*.

> *VIDEO*: African native dancing is merged with current rock and disco dancing. Dissolve to a brief choral climax from Mahler's Eighth Symphony.
>
> Dissolve to a Frenchman sitting on the garden bench with a magnificent chateau in the background.

FRENCHMAN: The joy of appreciating Mozart or Molière or a Michelin three-star restaurant is greater than the pleasure from rock music or McDonalds. But it requires education. Civilized tastes are artificial; they are acquired tastes. To become a golden age of civilization we need education for civilization.

A civilized education was a great rarity in the second half of the twentieth century. A civilized education teaches us how to enjoy things.

In the twentieth century, educators distinguished between a liberal education and a scientific education. A person took either the arts or the sciences. A scientific education taught how to build things or do research. A liberal education taught how to be a lawyer or diplomat or executive. Both were preparation for how to work or make money rather than how to enjoy life. Both concentrated on *means* rather than *ends*.

A civilized education (which existed in many earlier societies) is designed for a more leisured class—for people who do not have to become managers or engineers or have their nose to a mental or economic grindstone.

Robot factories and office automation took us, screaming and kicking, into an age of greater leisure. An age of leisure needs education for leisure—education that develops the exquisite pleasures of appreciating great books, great art, great computer software, and great music.

VIDEO: A conversation between the narrator and a university professor while standing on a moving sidewalk in a traffic-free city precinct. They are carried past sculptures, carefully lit for the appreciation of the people traveling on the moving belt. They are exact reproductions of sculpture from Athens. Among them are richly decorated friezes and Corinthian columns.

UNIVERSITY PROFESSOR: Of all societies known to history, it is usually agreed that the one in which the numerous facets of civilization came together most acutely was that of Athens from about 480 B.C. to 323 B.C. The Athenians of the fifth century B.C. were probably the first society to train themselves deliberately for the appreciation of life—for the subtle and intense pleasures of erudite debate, witty conversation, athletics, beautiful buildings and sculpture, and articulate theater.

This is almost the opposite of the traditional ethic of Hollywood and Madison Avenue, which says "Nobody ever lost money by underestimating the taste of the public."

You ask what it means to be civilized. In my view it relates to the deliberate development of the mind to enjoy the most subtle pleasures. You find this characteristic in all of the golden ages of civilization: the self-conscious training to enjoy and perfect music, art, theater.

NARRATOR: And poetry, literature, conversation, food, wine, flirtation . . .

PROFESSOR: Yes.

NARRATOR: And television, film, news commentary, parks, architecture, the environment . . .

PROFESSOR: Today, yes. The enjoyment and good states of mind that come from full appreciation of these diverse arts are the hallmark of a great civilization. A great civilization trains itself to make the most of its powers of thinking and feeling.

NARRATOR: Today this includes making the most of our computers and optical disks to enhance our thought processes, expand our knowledge, eliminate drudgery, and improve the creativity of composers, sculptors, architects, and television makers. We use whatever *means* are appropriate to achieve the *end* of enriching life to the fullest.

PROFESSOR: The Spartans discovered that a whole community could train itself for war. America discovered that a whole community could train itself to make money. We are now in a process of discovering that a community can train itself to be civilized and that a higher level of joy comes from that than from immediate and easy gratification of instincts.

NEWS COMMENTATOR: Project Jacob as presently conceived is the wrong engineering. The orbital rings idea is not right technically. It is too expensive and difficult compared with a cable directly to geosynchronous orbit. We cannot build that cable yet because we need cables with ten times the tensile strength of today. But we will develop them.

It is necessary to get the timing right for great advances in technology. We should build the space catapult now; this will teach us much about cables in space. But stages 2 and 3 should be shelved and the money spent on research so that geosynchronous elevators can eventually be built. That is the simple and elegant route into space. Great engineering finds the simple solutions.

HEAD OF PROJECT JACOB: If we wait for technology to improve, we may wait indefinitely. It will become an academic scheme as opposed to a real project. The building of the tunnel between France and Britain waited 150 years. *Now* is the time to start this endeavor.

The economic benefits are immense. Project Jacob is similar to the building of the railroads to California in the nineteenth century. That project too looked outrageously expensive and difficult. It took a higher proportion of America's gross national product than Project Jacob will take of the world's gross national product. But it opened up a new era in human history. Jacob will build railroads into space. It will open up a far more glorious era in our history.

PROFESSOR: Who needs railroads into space? Our problems are here on earth. Our objectives should be to improve the quality of life on earth.

HEAD OF PROJECT JACOB: Human activities in space have added greatly to the quality of life on earth—the manufacturing of valuable pharmaceuticals, the growth of exotic crystals and components for microelectronics, the creation of special alloys, telecommunications, earth surveillance for crop growing, weather forecasting and mineral prospecting, and particularly the broadcasting of many high-resolution TV channels.

It is clear that there would be much more activity in space and that major space colonies would develop if the cost of transporting people and materials to and from space were not so expensive.

PROFESSOR: The cost of Project Jacob is totally outrageous. It's a hysterical reaction to the Russian expedition to Mars. For the same price you could give a million artists and sculptors a grant for many years for improving their skills. If you want to improve the quality of life, spend money directly on the quality of life, not on railroads into space!

HEAD OF PROJECT JACOB: It's nice to have art and sculpture. We have vastly more now than 40 years ago. The world is a better place for it. But civilization is more than that. New Yorkers at the end of the nineteenth century could have said, "Don't build railroads to California. Spend the money on art instead."

PROFESSOR: New York would have been a better place!

Athens was a society rooted in a constant vigorous, competitive pursuit of excellence—excellence in athletics, in the arts, in intellect, and in debate. It was not a rich society. By today's standards it was financially poor. But it had wonderful architecture, sculpture, drama, and dinner parties. It was an immensely creative society that encouraged originality, a society that praised excellence in all spheres and did not distinguish between the spheres, a society that could not distinguish between art and science because it had only one word for them both.

Athens is often quoted as a paragon of democracy. Sovereignty was vested not in a king, a parliament, or a president but in the assembly of all citizens. The citizens had more direct say in their own government than in any society since.

NARRATOR: The democratic freedom applied to citizens only, and these were a fairly small portion of the population. Women and slaves did not participate in the assembly. Today's electronic participation in the debates about government is intended to apply to everyone. Only the machines are slaves.

MUSEUM CURATOR: If I were to choose a golden age of the past to live in, I would prefer Renaissance Italy to Athens.

VIDEO: A montage of the churches and works of art of the Renaissance.

CURATOR: The ideas were more advanced, more complex, the sculpture and painting more beautiful.

VIDEO: The camera explores works of Leonardo da Vinci, showing their amazing diversity, lingering on beautiful details, dissolving from faces to engineering sketches to architecture to anatomical drawings.

FRENCHMAN: (again in the chateau gardens) And If I were to choose, I would select the Paris of Louis XV.

VIDEO: The camera shows magnificent chateaux, their formal flower beds with geometrically shaped paths, sculpture on the buildings. It wanders over the painted ceilings and dissolves to an engraving of a grand ball. It shows details of the participants, then a painting of Madame de Pompadour in her finery.

FRENCHMAN: Again, the ideas were more advanced—the debates and writings of Voltaire, Montesquieu, the *salons*, the Parisian society, the birth of the Age of Reason.

AMERICAN: If I could choose an age to live in, it would be today. Today's ideas are far more complex and interesting. We have access to vast libraries of books and videodisks. For the first time the writings, images, plays, and videos of all human history can be at our fingertips. We do not have the bloodshed and torture of the Renaissance. Eighteen-century France culminated in one of the bloodiest revolutions in history. Each successive golden age ought to be better than the previous one. The ideas progress, knowledge accumulates, our tools and media improve, the storage of human culture becomes larger. We haven't got our act together fully yet, but if we can imbue this world of advanced technology with the right sense of values, it can become far greater than the great periods of the past.

NARRATOR: Today, as we define what we want society to be, we can demand characteristics that were unavailable to the golden ages of the past. Our automated factories and robot machinery in the home give us wealth and leisure. We have time, if we wish, to learn to appreciate the most refined pleasures of an advanced culture. We have superlative music reproduction. We have worldwide communication. We have access with our wall screens and public videodisk libraries to the finest drama, documentaries, education, art, and debate. There is no need for slaves, physical or mental poverty, drudgery, or moronic bureaucracy. Intellectual debates and inventions are greatly enhanced by the vast knowledge that computers process.

There *is* need for much thought about what we are doing with technology. It is clear that we are unlocking technology that to an earlier era would have appeared like magic, but we are naive about what to do with it.

> *VIDEO:* (over "Sorcerer's Apprentice" music) Montage of advanced technology. A microscope shot of a needle piercing the walls of a cell. An X-ray laser weapon in orbit. A robot creating sculpture. A scene in an opera house showing a science fiction version of Wagner's Ring Cycle. Fast camera movement through a multitiered hydroponic greenhouse with strange plants. Microscope shots of a complex cell dividing. Slow dissolve to an image of Aladdin's lamp and its genie.

NARRATOR: This is still only the beginning. It seems clear that technical advances in the next 40 years will be far greater than those since 1980. Yet we do not understand the meaning and purpose of it all. Technology has given us the capability to work miracles. How do we build the most civilized society we can in such a world?

We cannot do so without a clearly expressed system of values. The values act like genes in establishing behavior patterns in a highly complex organism. If the genes are faulty, the organism will go astray.

FRENCHMAN: (in the chateau gardens) It is ironic to reflect how Europe used to worry about the impact of technology as it watched the high-tech shoot-out between America and Japan. In fact, Europe was much closer to the sense of values needed for a highly developed society than either America or Japan.

> *VIDEO:* We observe the robots tending the chateau garden and see briefly the controls for the hydroponic flower beds.

NARRATOR: Earlier in the series we described the values of society with a set of four blocks.

VIDEO: The four blocks—"Basic Ethics," "Political Constitution," "Welfare System," and "Civilization Values," as in Figs. 1.1 through 1.4—are built up on the screen.

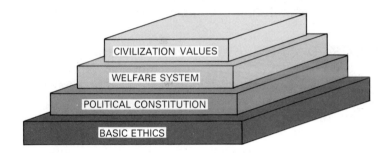

NARRATOR: There is general agreement about the bottom three blocks. The bottom block is the value system collectively expressed by the world's religions and now taught thoroughly in schools everywhere, usually independently of any one religion. The second block is the constitution defining democracy and the rights of humankind. The construction is also taught to schoolchildren and should be fully upheld by the media. The third block relates to the basic necessities of life. We need to make sure that all members of a society have adequate food, housing, education, medical care, police protection, and opportunities to work. The nation as a whole needs good defense.

The bottom three blocks are the basis of laws. They are the foundation stones of any civilized society today. The top block is more subtle and relatively recent. Whereas the lower blocks are non-controversial, the top block causes endless debate. The mere existence of that debate is vital. A society can be fully democratic, and its citizens may have total freedom and substantial wealth through automation, but it can still be a barbaric society. Its universities, like many universities in the recent past, may be turning out barbarians.

VIDEO: We see a section of the city with sculptors' studios and houses that are partially underground. Parks and gardens are landscaped on top of the buildings. Colorfully dressed people stroll past blooming rhododendron bushes in flower. A group is practicing a Mozart clarinet concerto.

SOCIAL WORKER: The purpose of life surely should be more than personal enjoyment, even if the level of appreciation is raised to highly civilized planes. The goal of a good life should be that we leave the world in some way a better place than we enter it.

NARRATOR: A worthy goal. But what do you mean by "better"? The debate about civilization asks, What is a better society?

SOCIAL WORKER: A society without poverty. A society that is safe for its citizens, where everyone is well fed and well educated. A society free from pollution. A compassionate society where sickness is minimized.

NARRATOR: Those are important basic objectives, and they have been achieved to a large extent. You cannot go on to higher-level goals until you have dealt with poverty and hunger and sickness. But then what? Suppose that society is clean, safe, well fed, rich, and healthy. It could be a soul-destroying, clinical place where most people are bored and listless.

CYNICAL AUSTRALIAN: (in a screen window behind the narrator) In most technological societies we seem to be creating something closer to Huxley's *Brave New World*. The alphas and the epsilons. An educated class and a slob class. Unlimited opportunities for the educated and the adaptable, endless pap for the stupid.

Advanced societies have eliminated poverty but have not eliminated slobs. The first is the prerequisite to the second. The second may not be possible, but it's worth trying.

NARRATOR: It's really completely different from Huxley's vision. By the year 2000 the term *working class* had disappeared in many wealthy societies. The more arrogant media were using the term *slob class*. Choice of language can have a strong influence. Many young people strove not to be part of the slob class, just as with earlier language many had aspired to be "yuppies." But after a time the slob class became fashionable. There was slob rock, slob decorators, a slob channel on cable TV. Slob fashions became a $10 billion industry. The rebels against "civilized" values made themselves heard at maximum amplitude.

This is entirely different from Huxley's *Brave New World*. Slob fashions are an expensive deliberate choice—a barbaric choice, perhaps, but it should not be a surprising choice given the top-block value system of many schools and universities.

BRITISH PSYCHOLOGIST: The really true advances in civilization are those that expand the mind. It has become very clear that the brain of nearly everybody is capable of far greater achievements. Societies of the past have failed to develop the brain. Many of today's institutions cause severe damage to our ability to use our brain well.

VIDEO: A montage of pictures of peasants from seventeenth-century paintings of Brueghel. The sequence shows close-ups of their faces, which suggest the total absence of culture, literacy, education.

PSYCHOLOGIST: These peasants had the same physical brain as ourselves, but it was never trained, never taught to be creative, to have intelligent conversations.

VIDEO: Film of the deadened faces of workers on a car production line in the 1970s, deafened and dirtied into stupidity.

PSYCHOLOGIST: These people have brain damage caused by 20 years of moronic work. The attempts to retrain them at age 40 to do useful, enjoyable jobs was largely a failure.

VIDEO: Film of bureaucrats talking with refined accents at a London club.

PSYCHOLOGIST: These men all got first- or second-class honors at Oxford or Cambridge—the cream of their country. By 35 they are so set into rigid patterns of behavior that they have largely lost the ability to think in other ways.

We now know much more about the human brain than we did in the old days when IQ tests were our main measure of human intelligence. A young child has many potential intelligences, centered in different portions of the brain—linguistic intelligence, musical intelligence, mathematical and logical intelligence, spacial and visual intelligence, and so on. Different children have strengths in different areas. These need to be detected early and developed to their full potential.

VIDEO: Film of 3-year-old children in Japan having violin classes. A Vivaldi quartet being played by 6-year-olds.

PSYCHOLOGIST: When the functions of the various modules of the brain are not developed and trained in a young child, the potential of the brain for using those forms of intelligence is lost. In most children most of the potential was indeed lost until relatively recently. Even today it is only a minority of parents who understand the ideas of multiple intelligences in the brain and who detect and develop those areas in which their child happens to excel. A civilized society must surely be one that develops rather than loses the potentials of the brain.

VIDEO: Dissolve from a montage of Suzuki-style music classes and excited faces of Japanese children to Western children racing down the streets of a city, neglected by their parents.

PSYCHOLOGIST: Later in life the specialized functional areas of the brain lose their ability if not exercised. In the age of drudgery, before automation, most people lost most of the capability of their brain through being chained to forms of work that trivialized and exhausted them. By age 40 irreparable harm had been done to them. Often we needed to retrain them for fundamentally different jobs because of changes in technology, and the retraining proved to be almost impossible.

NEWSREEL: Sequence of scenes from the 1984 coal miner's strike in Britain. Coal miners, finally back at work, leaving the pit head with miserable faces blackened with coal dust.

PSYCHOLOGIST: Coal mining 40 years ago must have been one of the world's worst jobs. Filthy, moronic, bestial, utter human degradation. These miners were out of work for a year, on strike due to the necessary closure of unproductive pits. Their future in mining was desolate, but almost none of them opted for retraining in cleaner jobs.

Many big corporations forcefully retired high-pressure executives at 60. The statistics show that many of them died within five years, unable to adapt to productive new activities. Many people with 40 productive years ahead of them have lost the ability to learn.

VIDEO: Dissolve to group of dons at Oxford, all over 75, wearing the gowns and regalia of the university.

PSYCHOLOGIST: Men who spend their life learning and researching are able to do it fast and competently into old age. They keep the learning circuits of the brain exercised.

A civilized society must surely be one that avoids brain damage through drudgery or neglect. Old people, when their physical energy declines, should still have mental circuits that are active, fully developed and exercised, and they should be respected for the wisdom that accumulates throughout life.

VIDEO: An animated lecture being given by an 85-year-old Chinese professor. Dissolve to a 90-year-old conductor, his face alive with excitement and professional precision as he controls the great music.

CURATOR OF THE RIJKSMUSEUM IN THE NETHERLANDS:
The vast majority of thinkers and philosophers who respect the great
civilizations of the past would say that technology is almost totally
irrelevant to the goals of being highly civilized.

The great Dutch painters did not need technology, and today,
in an age of technology, nobody can paint like them. If Michelangelo
had lived today, he could not set the standards for a culture as in
Renaissance Italy; he would barely be noticed in the rabble. If Shakespeare were alive, he would be writing scripts that would be massacred
by some director under corporate pressure.

NARRATOR: Shakespeare had the genius to adapt his work to the
popular media of his day. I suspect he would have done the same
if he were living now.

CURATOR: Shakespeare did not need computers and video. Nor
did Leonardo. Nor did St. Francis.

NARRATOR: No society can guarantee producing saints or Leonar-
dos. Shakespeare was one man in a millenium. What about all the
other clever people in Elizabethan England who never had the chance?
We have a vastly greater probability of creating geniuses now. All
schoolchildren are educated for creativity. They don't have to indulge
in the drudgery of earlier decades.

CURATOR: (with wit in his eyes) And they created sewage jazz
and flasher dancing.

NARRATOR: They also created the hologram museum and the best
choral music in history and the Melville Island Bridge, and Stobart's
films of the Ring cycle.

CURATOR: Renaissance Italy needed no technology, nor did the
great era of France or Athens. We might ask whether the influence
of technology is good, bad, or neutral. One would have to be very
generous to say that it is neutral. Its main effects are endless garbage
on TV, the terrors of future war, mind-destroying pressures in the
work environment, traffic changing the cities from pleasant places
into polluted, noisy hellholes.

NARRATOR: That sounds like twentieth-century technology. Surely
the whole point of technology today is to build a better world. If
we don't build a better world with the riches we are unlocking, our
value system has failed.

There are innumerable examples of an early use of technology
causing bad side effects, sometimes very bad, and a more advanced
technology correcting the bad effects.

VIDEO: A sequence of images follows. The first of each illustrates bad use of technology; the second shows good use.

We see a city traffic jam, fumes rising from cars revving their engines. Pedestrians on a narrow sidewalk can barely talk for the noise. When the traffic moves fast, pedestrians run across the streets in front of it. This is contrasted with a city where the traffic is fully separated from the pedestrian areas. The pedestrian areas have parks, people movers, sculpture everywhere, outdoor cafés, interesting displays, orchestras, dancing, small groups gathered around orators. Such city areas have become an environment for people to interact.

We see the stupefied faces of the workers in an automobile assembly line and then the electronic ballet of the robot production line and the relaxed face of a woman at the console monitoring the robots.

We see how television changed elections. We see some inanities mouthed by politicians of the early 1980s: the Carter-Reagan election broadcasts; Coluche the clown contending in the presidential campaign in France, in a way less ridiculous than the real politicians. This is contrasted with images of participatory democracy in which the public interacts on their keypads with skilled TV interviews on political issues.

We see a typical derelict industrial landscape of 1980 or earlier: slag heaps, broken machinery, abandoned buildings. We see the same landscape converted into an interesting park, the slag heaps now green hills with trees chosen for their beauty. Children play around the ponds and climb on the old machinery, which has become like park sculpture.

We see a montage of the throwaway society of the 1970s and 1980s: the rubbish discarded by an average family, the waste of food, the waste of materials. The daily amount of good, edible food thrown into the garbage by one American family. The used-car graveyards. This is contrasted with machines sifting every bit of waste for recycling.

Cut to a conversation in a mass-production workshop where computer-controlled lasers are replicating sculpture.

SCIENTIST: (in a white laboratory coat) Particularly important to progress is the preservation of our knowledge in a computer-usable form. Old people who have used their brain fully have immense knowledge. Regardless of what we would like to think about an afterlife, when we die the capability and skill so painstakingly developed in the brain is lost forever. A civilized society should capture the unique skills and expertise. They should be organized into knowledge bases that will benefit humanity from then on. Great performances on organs or pianos should be filed in computers so that electronics can play the instrument exactly as the human performer did. From now on we have the ability to preserve and index the brilliance of individuals. The electronics and optical disk libraries are a vast accumulator of human genius.

VIDEO: The camera wanders lovingly over half-complete replicas of Michelangelo sculptures, with brightly lit smoke from the laser cutter.

PROFESSOR: Technology is enormously important but must be subject to the value system. In order to be civilized we need wealth, and technology creates wealth. When we describe Louis XV's France as a great era in civilization, we are referring to a tiny proportion of the population who had vast wealth. It took an enormous amount of money to build the great chateaux and parks. The artists, as in Michelangelo's day, needed *very* wealthy patrons, such as the Catholic Church.

Democracy as we define it today—democracy for *all* the people rather than for a minority—needs fairly advanced technology. There must be no slaves and no enforced poverty. People should not lead lives of mindless drudgery. There needs to be good education and responsible use of media. Democracy works much better if there is interactive television. Most attempts to establish democracy in poor, struggling, or technically backward countries have failed. There have been few long-lasting democracies in the third world. Countries that have climbed out of third-world status have often established the basis of a democracy that appears to be stable.

NARRATOR: The debate about whether technology is good, evil, or morally neutral is an ancient one. It surfaces frequently with the viewpoints of different philosophers. The argument is like a theological one. Throughout history religious philosophers have often been negative about technology. The pope made a recent broadcast on the subject.

NEWSREEL: Video footage is shown of Pope John Paul V.

NARRATOR: John Paul V was elected pope two years ago. He is the first African pope. He was born in Uganda in 1965, suffered appalling starvation under Idi Amin and in the civil war that followed, studied biochemistry at Harvard and then theology, and continued to publish scientific papers up to the time he was elected cardinal. He is a saintly man, with a powerful gift of oratory. After his first visit to Australia this year, he made a worldwide broadcast.

VIDEO: We see the scenes described intercut with the pope speaking direct to camera with a video window behind him.

The pope starts by presenting a series of technological healings: artificial limbs, artificial intelligence by which a nonspeaking spastic can communicate, electrodes implanted in the skull so that a deaf person can hear, direct action on the visual cortex so that a blind person can see, and so on.

This sequence states a powerful case for technology. People whose life functions would otherwise be minimal can now lead almost normal lives.

> *Pope*: How can this not be a great good? The blind see, the deaf hear, the lame walk.

He then speaks about the historic failure of the church to claim science as its own. He presents the meeting between Clement IV and the thirteenth-century scientist Roger Bacon, who believed that scientific knowledge was divinely inspired. He tells how, regrettably, Clement died too soon, and the church followed St. Thomas Aquinas, who asserted that scientific knowledge could not be acquired without divine inspiration, and thus prepared the ground for the split between theology and science that came to a head with Galileo. Pope John Paul V declares Bacon right and St. Thomas wrong.

The pope creates a vision like that of Teilhard de Chardin—the unifying of human consciousness, the world as one. All peoples linked and understanding one another through God. Technology is eliminating barriers between peoples. The world can be united. Not dully united in one blurred obliteration of individuality. No, the world is only one because it is a world in which each individual is fully himself, a world of such vibrant individuality as history has not yet conceived. Each person in the world contributes as an expert to world civilization because he is indeed an expert—on his own uniqueness.

The pope makes an impassioned plea for technology to be applied to feed and heal the hungry of the third world as in his own country of Uganda. Only when poverty and hunger are abolished can war be abolished. He calls for prayer in all nations, unified, on world television, to avoid the dangers of war and to eliminate the weapons of war.

Science can be the tool of the Devil or can be used for the work of God. It must be used for God's work, and that needs God's guidance. Technology has become awesomely powerful. Used wrongly it will wreak havoc on a scale unknown. In an age of advanced technology man needs prayer even more than in earlier days. It is only by deeply probing our conscience and asking for God's guidance that we can know how to use technology for good and not evil.

The four building blocks return.

NARRATOR: (back in the restaurant in Paris) So what should be in the top block of our value system, "Civilization Values"? It can be different from the Athens of Pericles or the France of Voltaire because today we have the technology to improve upon those.

What, then, is a civilized society?

- A society in which a highly developed sense of excellence predominates in the value system, excellence in art, literature, drama, music, debate, sport, science, technology, architecture, the environment, television, dinner conversation. A society in which the sense of excellence is taught to children and pervades the media.

- A society in which reason dominates. No subject is taboo. Rational and skilled debate is perceived as a major form of enjoyment. Excellence in articulation is cultivated. All dogma is open to question.

- A society that develops the numerous forms of intelligence in its children as fully as possible; a society that prevents brain ossification as its citizens mature.

- An open society, allowing maximum personal freedom and encouraging each citizen to develop his or her individual talents and capacities to the fullest while remaining tolerant of the individuality of others.

- A society so confident of its traditions that it is able to encourage constant experimentation to improve itself.

- A society in which every individual feels that he or she has a part to play. No class should be unable to contribute.

- A society that makes connections—between programmers and poets, between physicists and filmmakers, between doctors and clowns, between engineers and tennis players—so that the skills of one craft can be assimilated by the other craft to their mutual enrichment.

- A society run by its members—all of its members.

- A society blind to trivial differences like sex, color, nationality, or faith, which concentrates instead on the deep values and abilities of people.

- A society that exults in the diversity of human personality.

- A society with sufficient leisure that its citizens can learn to appreciate great music, drama, art, debate, and literature.

- A society that minimizes drudgery, abolishes bureaucracy, and avoids forms of work that captivate the body or mind to the exclusion of civilized pursuits.

- A society that cultivates a sense of humor and gracious manners. A society that can laugh at itself.

- A society where the environment is beautiful, unpolluted, and appreciated as an art form.

- A society that captures human excellence and expertise digitally so that it builds ever greater storehouses of knowledge, art, and machine intelligence, enabling future generations to build upon the achievements of previous generations.

HEAD OF PROJECT JACOB: This is fine, but humankind needs more. It needs adventures, grand adventures. It needs new frontiers to conquer. Civilization needs challenges on a grand scale, and these should be international challenges. If we do not have challenges that stretch us to the limit, we will probably drift into war situations between major powers.

Where there is a frontier land, it will be tamed. Where there are barriers, they will be broken. If our *only* challenges are to make art and television, we will become weak and effete. We will become museum creatures. Space today is the new frontier, the great challenge to our age. If we accept this challenge, we shall find adventure, new knowledge, excitement, fulfillment in subduing space.

In the twentieth century we lost the vision to manage projects of long duration. It took two centuries to build St. Peter's in Rome. We need once again to have grand visions, projects that span lifetimes. Civilization, I agree, should mean the striving for excellence—but more than just excellence in literature, arts, and sensual satisfaction. We need great projects, goals that expand enormously human capability. We have had intelligent machines for a time span that is nothing in evolutionary terms. We need visions of how we evolve.

You have commented that the technical advances of the next 40 years will be much greater than those of the last 40. This acceleration will go on for a millennium.

Computer intelligence will become greater and greater. The knowledge in computers never dies, unlike that in our brains. Our ability to recode genes will improve and improve with genetic engineering processes being controlled by computers. Computers themselves will use biological circuits. Our unlocking of subatomic particles will be a billion times more interesting than the splitting of the atom, and probably more dangerous.

Our greatest engineering accomplishments will be in space. We will soon have a long-chain-molecule fiber so fine and strong that many thousands of miles of it can be coiled up on a large drum, taken to geosynchronous orbit, attached to a space vehicle, and strung across space. It will become more and more possible to shift everything that is dirty and dangerous to space and concentrate on making the earth a beautiful place.

Our value system must strive for greatness in everything, but above all for greatness in extending the frontiers of humankind.

REFERENCES

1. Genesis 28:12.
2. Paul Birch, *Partial Orbital Ring Systems*. Washington, DC: NASA, 1981.
3. Paul Birch, *Space Exploitation Using Jacob's Ladder to Replace Launch Vehicles*. Washington, DC: NASA, 1980.

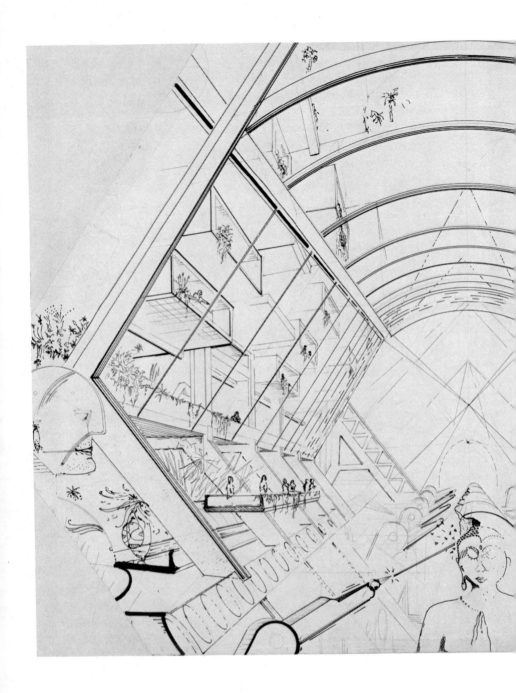